KEEPER OF SECRETS

CARLY MOORE MYSTERY

BOOK SEVEN

DENISE GROVER SWANK

ALSO BY DENISE GROVER SWANK

Carly Moore Mystery

Complete series that begins with:

A Cry in the Dark

Maddie Baker Mystery

Blind Bake

Bake Off

Rose Gardner Mysteries

Multi-book series but start with this one:

TWENTY-EIGHT AND A HALF WISHES

Magnolia Steele Mystery

Center Stage

Act Two

Call Back

Curtain Call

Darling Investigations

Deadly Summer

Blazing Summer

Created with Vellum

CHAPTER ONE

MARCO

"I'm not runnin', Marco," Max yelled into my ear through my cordless landline phone. Cell phones didn't work out here in the Appalachian Mountains.

"It's not *running*, Max," I insisted for what felt like the tenth time. I was pacing in my bedroom, trying to convince my best friend to leave our hometown of Drum, Tennessee with me and my girlfriend Carly. We needed to find proof that Carly's father had not only killed her mother but also wanted Carly dead as well. We needed to neutralize him and the threat he posed.

First, we were stopping at a jail in Arkansas, hoping to visit a prisoner who'd supposedly worked with the asshole. I was sure it was a pipe dream. For one thing, James "Skeeter" Malcolm had absolutely no reason to agree to see us, and even if we got in, the chances of him telling us anything useful were slim to none. Still, Carly's friends in Arkansas knew Malcolm. We were hoping they could convince him to see us.

Part of the reason I wanted Max to come with us was to keep him safe from his brother Wyatt, whom we'd just discovered had been keeping notes on just about everyone in town. We'd stolen those notes, and when he found out, he was likely to go on a rampage. Considering he had shackles embedded in his basement floor, I didn't want anyone I loved to get caught in the crossfire.

But Max wasn't having it. After a lifetime of shying away from conflict and acquiescing to his family, he was ready to stand up to them. While I was proud of him, this wasn't the way to do it. Not now. Not when his brother had run off the rails and saw all of us as threats.

"It's not running. It's self-preservation," I said, "and you know we could use your help. *Carly* could use your help."

"My family's in chaos," he said, sounding exhausted. "My father's been arrested. My mother's likely fled the country, and my brother—"

"Your brother is *dangerous,* Max."

"He won't hurt me," he ground out.

"Maybe not before," I said, my anxiety building. "Before we discovered his secrets."

A loud thud sounded from the other side of my closed bedroom door, and my heart lurched.

"Son-of-a-bitch!" Hank called out in panic. Frankly, I still struggled to believe that Hank Chalmers, the former pot king of eastern Tennessee, was not only in my house but here on friendly terms. I'd heard stories about him growing up, and even more of them as a deputy sheriff, and it had never ceased to amaze me that he'd never been put away for anything. But Carly had befriended him within twenty-four hours of getting stuck in Drum eight months ago, and now he was like a father to her. He was good to her, and that superseded any natural acrimony there might be between cop and criminal. "*Marco!*"

My heart started racing. Hank was a stoic man who was as tough as shoe leather. There was only one thing that would drive him to panic, and that was if he was frightened for Carly.

"Max, I gotta go," I said, rushing for the door. "Something's happened to Carly."

"Don't hang up!" he shouted. "Tell me what's going on."

Somehow I had the wits to put him on speakerphone as I opened the door and bounded down the short hall.

Hank was grabbing a shotgun off the coffee table, balancing on his crutch as he bent over on his one leg. "The laptop's gone," he grunted without looking up.

"What?" I asked in confusion and disbelief.

"Someone took the damn laptop with all of Wyatt's files." He stood upright and headed for the door.

"Carly," I whispered as the breath was sucked from my lungs. Had she left and taken the laptop with her?

"Paper files are still here," Hank said, opening the front door as he shot a backward glance to the phone in my hand. "Suggests they were in a hurry."

As a sheriff's deputy, I'd been trained to handle high-stress situations, but I felt like I'd been submerged in icy water and was moving in slow motion, ten paces behind Hank as he damn near ran out the door into the warm summer evening. Except I was only a couple of paces back. What I was feeling was my brain slogging through mud.

"*Carly?*" Max asked over the speaker. "Do you think *she* took it and went after her father on her own?"

She would do exactly that to protect the three of us, but ten minutes ago, she hadn't given a single indication that was her intention. We'd come up with a plan for the three of us—four if we could convince Hank—to head to Arkansas within the hour. We were supposed to do this together.

"No," Hank barked, his gaze landing on my SUV, parked about fifteen feet from the house. "She might have pulled that stunt before, but not now, not after what we've learned over the past few days. She knows she doesn't have to do this alone. She knows there's power in numbers." He nodded toward the other vehicles parked in front of my house. "Carly's car and your Explorer are still here. She didn't leave." His voice turned gruff. "Not on her own."

It took a second for his insinuation to sink in, but then I noticed the open hatch on the back of my SUV. A speck of hope lit inside of me. Maybe she was packing the car.

"Carly?" I called out as I ran off the porch, heading for the vehicles. *"Carly!"*

"Marco, don't be a fool!" Hank shouted. "You need a firearm."

I knew he was right. Wyatt was a dangerous man, and he was worked up about the stolen files. There was a chance he was in the trees, waiting to retaliate with an ambush. My home was on a wooded lot with plenty of hiding places.

But fear for Carly's safety overwhelmed all reason. I *had* to find her. I had to make sure she was okay.

"Carly!" I shouted, running to the back of my Explorer, which was parked slightly perpendicular to the house, the back mostly hidden from the porch. Carly's suitcase was sitting on the ground as though she'd brought it out but had left it for me to put in for her. Only Carly was too independent for that. She would have done it herself.

"What's goin' on, Marco?" Max asked, his voice coming from the phone still clutched in my hand. I'd completely forgotten I was still holding it.

"Uh..." I ran my free hand over my head. "I don't know. My Explorer hatch is open, and her suitcase is on the ground in front of the back, but she's not here."

"Don't move," Hank snapped as he hobbled toward me faster than you'd think an elderly one-legged man could. Only he was making a broad sweep around the driveway instead of using the most direct path. "You're screwin' up the prints."

"Prints?"

"The damned footprints, you fool." He stopped and raised the end of his crutch, pointing it at me. "Don't you fucking move."

"What's going on?" Max repeated, more insistent this time.

I stood in place, my entire body shaking with fear and adrenaline. Hank was right. There were footprints on the asphalt around the SUV—larger prints made from dirt that looked like they came from a man's work boots, not Carly.

Hank walked closer and then leaned over to observe the ground. He and I may have both grown up in the Tennessee mountains, but he was more of a mountain man than I could ever claim to be.

"A man was here," Hank said, following next to the prints as they made their way toward the front porch, then faded. "The way his footsteps are spaced, he was leading her toward the house."

"Wyatt," I said in a harsh whisper.

Hank turned and headed back toward the SUV, then moved past it, keeping his gaze on the ground. The two-car asphalt driveway narrowed to a single lane to join the country road, and Hank's attention was focused on the narrower portion behind the Explorer. He stopped and stared at a damp spot on the ground.

"Carly's footprint is here." He pointed to the ground, and I joined him, looking over his shoulder.

A smaller print was embedded in the dirt over part of a man's shoe print.

"She was heading toward the road."

"Wyatt kidnapped her?" I forced out.

Hank didn't answer. He just kept walking along the driveway, then into the grass as the ground angled toward the county road. "He was dragging her here. The heel marks are deeper, so you can see she's resisting."

I felt like I was going to throw up.

"Jesus," Max said, sounding breathless. "I can't believe he would do that."

"Why the hell not?" I shouted, lifting the phone closer to my face. "Look at everything else he's done!"

"Calm down!" Hank ordered, continuing to walk toward the road. "Yelling at Max won't help anything."

He was right, but I was furious, and I needed to take it out on someone. Wyatt's brother was a good scapegoat, even if it wasn't fair.

"The vegetation is too thick here to show any tracks," Hank said, his gaze on the ground, "but the grass is trampled. I think he put her in his vehicle, and she put up a fight."

The way he spouted it so matter-of-factly rubbed me the wrong way. "We're not tracking a damned raccoon, Hank. We're trying to find out what happened to Carly!"

He looked up at me, compassion filling his eyes. "Take a deep breath, boy. Carly needs you to keep it together."

"I'm coming over," Max said. "I'm headed downstairs now."

"No," I said. "We need to find Carly, and to do that we need to find Wyatt."

"He's right," Hank said. "I wasn't in the bathroom all that long, and we know he must've come inside to get the laptop, which means he maybe got a five-minute head start."

"How do you know Wyatt took her?" Max asked. "How do you know it wasn't her father?"

"He has a point," I said, feeling helpless and out of control. Carly had met her biological father that morning, and he lived in fear of his brother. So who's to say the asshole hadn't called Randall Blakely as soon as he left to tell him Carly was alive and well and had been in Atlanta? "But why would her father steal the laptop?"

"Good question," Hank said as he turned and headed back toward the house. "I know one way we might get some answers, but we're gonna cover all our bases."

I quickly fell into step behind him. "How?"

"I've got to grab something from inside, and then you and I are gonna head to Wyatt's house. There's little chance he's there, but we'd be foolish not to look. Max, you're gonna head out to that secret hideaway you stayed at with Lula last December when the three of you were hiding from your father."

"What do I do if he's there?" Max asked.

"Don't go in," Hank ordered. "While you think he won't hurt you, *I'm* not so certain. If he's there, leave and call my satellite phone. We'll handle it together."

"You have a satellite phone?" Max asked in disbelief. "Since when?"

"Since I got my money back," Hank grunted. "Now get back on task."

Max protested, "If he's there, it's gonna take too long to leave, call you, then go back. He's my brother. I can talk reason into him."

"No," Hank barked, clomping up the front steps. "You're going to tell us if he's there, and then we'll come up with a plan. Don't go in alone."

"What if it's Carly's dad who took her?" I asked, standing at the foot of the porch.

Hank grabbed the front doorknob and gave me a grim look. "That's what we're about to find out."

CHAPTER TWO

MARCO

"By checking Wyatt's properties first?" I asked as I followed him inside. "Do you really think he'd be stupid enough to take her to either of those places?"

"We're also checking his parents' house and his garage," Hank said with a dark look. "And no, I don't think he's at any of those places, but as I said, we'd be fools not to look."

I ran my hand over my head again. What was I thinking? Why hadn't we called the authorities? "We need to bring in the state police."

"Hold up," Hank said. "We don't want to do that just yet."

"Why the hell not?" I asked in dismay. "There's only three of us. They can cover a hell of a lot more territory."

"You know they won't take a missing person report seriously until she's been gone for longer than five minutes," Hank said. "If it is her dad, he's likely to kill her sooner than that. In fact," he paused, his eyes narrowing. "If it was him, why wouldn't he just kill her *here* and be done with it?"

Talking about Carly being dead formed a thick knot in my throat, making it hard to breathe. I coughed out a tortured sound.

"She's not dead, Marco," Max said insistently from the phone in my hand. "If whoever took her didn't kill her in your front yard, then they want something from her. She's alive."

"Yeah." I nodded to myself, trying to believe it.

"And if Carly's dad is behind this, maybe he wants her to marry that bozo she was engaged to before he offs her," Hank said.

I drew in a sharp breath. "Not helping."

"Tough shit," he grunted, picking up his messenger bag from the kitchen chair. "Ignoring the facts isn't going to help us find her. Now get your ass in the house and grab every weapon you have, and then we'll find out what's what. I need to make a few calls." He pulled a walkie-talkie-looking phone out of the bag and headed to the door again.

"Quit your gawking, and get to work," Hank barked when I didn't immediately start moving. "Be sure to get the bag you packed before she was taken. No sense wasting time coming back here. Whoever took her likely isn't hanging around Drum." Then, glancing at the phone in my hand, he added, "Max, you need to stay in town."

"I hadn't planned on leavin'," he grunted.

"Well, good," Hank said, opening the front door again. "I need you to keep an eye and ear out for your brother."

"And my mother," he said. "I'm not totally convinced she really left." She'd disappeared after learning her money laundering may have been discovered through a missing file that contained proof of her crimes.

Max's supposition surprised me, but I didn't stop to question him about it. I needed to get my shit together and find Carly.

Max hung up, and I headed to my room to start gathering my weapons.

Contrary to what Hank might think, I didn't have an arsenal—a rifle, and two handguns, one of which was my service weapon. Bringing my service handgun didn't seem like a good move, so I'd be down one handgun. But I suspected Hank had a stockpile of weapons that looked like they belonged in an armory. I knew he'd put a lot of effort into defending his drug empire back in the day.

I packed the rifle and all the ammo I had into my duffel bag, then grabbed my shoulder-strap gun holster and slipped it on, sliding my Sig into place.

Hank was still on the porch when I finished, so I dropped my bags onto the sofa and started gathering snack food. Who knew where we'd be going or what we'd be doing? Having something ready to eat could come in handy. I made sure to include some nuts and bananas for Hank to help manage his diabetes.

Shit. How were we going to handle his insulin? I was pretty sure it needed to be refrigerated.

I grabbed all the bags—including Hank's—and headed for the door.

When I got out onto the porch, Hank was leaning against a post at the bottom with his phone up to his ear. "Let me know if you see him or his F-150 or his tow truck," Hank barked. "And if Carly's with him, shoot Wyatt first and ask questions later." Then he hung up.

"Do I want to know what you were talkin' about?" I asked hesitantly as I stopped at the top of the stairs, the bag straps slung over my shoulders. As a sworn officer of the law, I couldn't condone such things, even if I agreed with him.

"Probably not."

"Maybe just tell me the legal parts."

He pondered me for a moment. "I know a lot of people in this county. A lot. And I know a few outside of it too. I've got eyes and ears everywhere looking for the both of 'em. There's no love lost between Wyatt Drummond and the people in my network, and now they know he might have taken something of value to me." His eyes darkened. "Add in the generous reward I'm offering, and if Wyatt so much as shits in a fifty-mile radius, I'm gonna know about it."

I nodded, then tromped down the stairs toward the Explorer. "I'd prefer him alive."

"Me too. Gettin' shot's too good for him."

There was a hard look in his eyes as he said it, and I wondered again if it was a mistake not to involve the state police or even possibly the FBI. Maybe we should have been working with the FBI from the beginning. Sure, Carly had insisted she didn't have any information to help build a case against her father's criminal organization, but maybe she knew more than she realized.

A year ago, I wouldn't have even considered keeping anything from the authorities. Now...

Carly worried that my integrity was slowly being chipped away by hiding her secrets, and that I'd resent her for it one day. I had to admit that I'd crossed lines that I never would have crossed before, but this wasn't the time for self-evaluation. I'd deal with my guilt later. Right now, I needed to find Carly, because if anything happened to her...

Panic twisted my insides. We *would* find her. We had to.

I tossed our bags into the back of my SUV, then closed the hatch and headed back to the porch. The front door was open and I saw Hank grabbing the folders and stuffing them into his messenger bag.

"How do you plan to find out if Randall Blakely has Carly?" I asked.

"That's on you, boy. You're gonna call Tiffany Olson. I'd do it myself, but she'd probably spit in my ear through the phone."

Carly and I had met with Tiffany, Carly's mother's best friend, a week ago. We'd hoped she might know something incriminating about Randall Blakely. While Tiffany had confirmed she'd always suspected him of killing his wife, and had even hired a private investigator, she'd backed off after being threatened by Blakely. Tiffany was the one who'd arranged for Carly to meet her birth father in Atlanta this morning.

I squared my shoulders. "If we tell her Carly's in danger, she might help us. I got the impression she genuinely does care for her. I'll call her, but we need to get to Wyatt's. We're wasting time."

"I got one of my buddies to go over and check it out," Hank said, slinging his bag over his shoulder. "He's gonna get back to me."

I ran my hand over my head again, knowing my hair had to be sticking up on end after the number of times I'd disturbed it, not that I gave a shit.

Hank held my gaze. "I say we head to Wyatt's garage and question Junior. That's a better bet."

I narrowed my gaze. "When you say *question...*"

"If Junior wasn't good people, I wouldn't let his wife clean my house," he said in disgust. "I want to ask him if he's noticed anything odd lately."

"Fine," I said, "but it's after six. Junior's likely not there. We'll have to track him down at home."

"The garage is on the way to his place, so we'll swing by to check," Hank said, heading toward my Explorer. "Now let's get goin'. You can make your call from the car."

"Good." I was eager to do *something* because every second we stayed here put one more second of distance between us and Carly.

CHAPTER THREE

MARCO

Wyatt's truck wasn't parked in Drummond Auto Repair's parking lot, not that I was surprised. I doubted he'd kidnap Carly, then head back to work. But Junior's car was there.

I parked in front of the building and hopped out, not waiting for Hank before I headed through the open door of the two-bay garage. A Chevy Malibu was up on a lift, and I could see Junior on a board underneath it.

"Junior," I called out. "I need to talk to you."

He rolled out from underneath the car and sat up, grease smeared across his cheek. He shot me a worried look, likely because of my tone and the fierce look that had to be plastered on my face. "What's up, Marco?"

Hank clomped in, his shotgun slung over his shoulder. "Your boss around?"

"I ain't seen Wyatt in two days," Junior grumbled as he wiped his hand on a rag from his coverall pocket. "We've got a bunch of cars lined up, waiting to be worked on. Donny Tucker's madder than a wet hornet 'cause Wyatt said he'd have his car ready by this morning, and I haven't even gotten to it yet."

"Where's he been?" I asked.

Junior narrowed his eyes. "Who's askin'? Marco, Wyatt's brother's best friend, or Deputy Marco Roland?"

"I'm askin' for *me*, Junior," I said, my voice breaking. "We think he took Carly."

His eyes widened and he clambered to his feet. "What?"

I gave him a pleading look. "What do you know—"

"When was the last time you saw him?" Hank barked, interrupting me.

Junior looked bewildered. "I haven't seen him since before Ruth's murder."

"When, exactly?" I asked.

"Well, I guess I last saw him the afternoon before," he admitted. He narrowed his gaze. "Why do you think he took Carly?"

"I need to see his office," Hank said, heading for the closed office door.

"It's locked," Junior protested. He made a move toward him and then hesitated, clearly torn between loyalty to his boss and his desire to help us.

"You got a key?" Hank asked.

"No—"

Hank lifted his crutch and smashed the end of it into the glass in the office door. When the reinforced glass didn't crack, he swung his shotgun around and pointed it at the glass.

"Wait!" Junior shouted. "You can't just go shooting up the place!"

Ignoring him, Hank unleashed a couple of rounds. The first cracked the glass in the corner above the doorknob, but the second shattered it. He walked over, grabbed a shop rag from a nearby shelf, and reached through the hole.

"What the hell are you doin', Hank?" Junior shouted in dismay.

"Lookin' for answers," Hank said as he unlocked the door then shoved it open.

"I'm sure if you call Wyatt, he'll…" His voice trailed off as Hank walked in and flipped on a light.

"You got the password to this thing?" Hank asked, dropping into the office chair. I could only assume he meant the computer in front of him. I should be the one in there looking. Hank had limited computer experience, but something Junior said stuck out.

I pulled out my cell phone to find Wyatt's number, then went to the front counter in the waiting room and called him from the phone. Instead of going straight to voicemail like cell phones usually did around here, the phone rang. He didn't answer, but the fact that it had rung meant he was somewhere in cell range.

"I can't get into this computer," Hank called out, his voice muffled. "Get in here."

"I'm tellin' you that you can't be goin' through Wyatt's things," Junior protested from the office doorway, but there wasn't much heat in his voice.

"Did you know Wyatt was spying on just about everyone in town?" I asked.

He turned back to stare at me, his mouth hanging open. "What are you talking about?"

"Wyatt was keepin' tabs on people and making notes in files," Hank said, giving Junior a hard stare. "There's even a file on you and your wife."

Junior stared at him in disbelief. "What kind of notes?"

"The spyin' kind," Hank said. "He's up to something, and he's pissed we know about it. He made plenty of threats against me, Marco, and Carly, and now Carly's gone missing."

"Someone kidnapped her out of my front yard," I said, my throat tight. "So you need to make a decision right here and now—do you want to be an accomplice to what Wyatt's up to, or will you help us find Carly?"

He stared at me in shock for several seconds before he responded, "That doesn't make any sense." He shook his head. "Why would he kidnap Carly? Is he holdin' her for ransom?"

"Shit," Hank grumbled, looking up at me.

Was that Wyatt's plan? Maybe it was plan B. When he'd called threatening Carly, he'd acted like a desperate man, and desperate men didn't act rationally. But if Wyatt had kidnapped her for monetary gain, why hadn't he left a ransom request?

Hank's satellite phone began to ring, and Hank reached into his bag and pulled it out.

I tried to contain my anxiety. It could be someone with news about Wyatt. Maybe they'd found Carly.

Hank pushed a button, then answered, "Chalmers."

"I'm at the cabin," Max said through the phone's speaker. "It looked empty, so I came inside. No one's been here for a while."

"We knew it was a long shot," Hank said.

"Okay," I said, "what about one of those tracking apps? Any chance you have one for Wyatt, Max?"

"You mean like Life 360 or Find My Friends?" Max released a dark chuckle. "What good would that do out here when we can't even use our phones to make calls? Even if it did work, there's no way in hell Wyatt would share his location with *me*."

Before Carly came to town last November, the brothers had barely spoken to each other in years. While they'd mended fences after discovering Lula was their half-sister, they had never been overly friendly. And the last month or so had left them particularly strained.

"Dammit," I grunted, starting to pace. "I figured it was worth asking."

Hank's head jerked up, and he reached into the front pocket of his button-down shirt. He held his non-satellite cell phone up to his face and then gestured to me. "Check this out."

"What?"

He shook his hand at me. "I have one of those apps installed on the phone I gave Carly last night. I damn near forgot about it."

I stared at him in disbelief. Did Carly know he could track her? But right now, I didn't give a shit about the morality of it if it helped us find her.

"I saw her put it in her pocket," Hank continued. "She had it when she walked out the door. She might still have it."

Hope flickered to life in my chest, but the phone had no service. "Wyatt's phone rang when I called him," I said slowly.

"If he's gettin' cell service," Junior said, "he's probably heading toward Buxton."

That made sense. While Drum was in a cell phone service abyss, Buxton, which was about ten miles north, was the closest town that got reliable cell service.

"Go!" Hank urged. "Junior here's gonna help me get into this computer. See what we can see."

"I'll head back into town," Max said over the phone. "Then we can figure out what to do next."

I gave Hank a long look, then bolted for the garage door and jumped into my SUV. I raced out of town and turned onto the highway that led to Buxton, making sure Hank's phone didn't go to sleep since I'd forgotten to get his password. About halfway to town, a few service bars appeared on Hank's phone. I pulled over to the side of the road and opened the location app, pulling up Carly's name.

A wheel spun for several long seconds. I was certain it wasn't going to locate her, but then it gave me something: Carly was north of Buxton, on a highway heading toward I-40.

I pulled back onto the road and continued in that direction, dialing Wyatt's garage as I drove.

"Drummond Auto Repair," Junior answered with a shaky voice.

"It's me, Marco. Let me talk to Hank."

"It's Marco," Junior said, presumably handing the phone over.

"I found her," I said in a rush. "She's on the highway heading north out of Buxton. Up toward I-40, it looks like. I'm following her now. I didn't wait for you because I don't want to lose her."

"That's good," Hank said. "After I made a call to Ginger, Junior here suddenly remembered the password to the computer. I'm doin' some diggin' now."

"I'll let you know when I find her."

"Good. As soon as Max gets here, we'll take off after you."

I started to tell him not to come, but if Carly's father was behind her abduction, I was sure he'd have multiple armed guards. I still hoped to involve law enforcement, but I wasn't ready to do it quite yet. I still doubted they would take this seriously, given she hadn't been gone all that long. "Yeah, sounds good."

"You be careful, Marco," Hank said in a husky voice. "And don't let anything happen to our girl."

"I'm gonna make sure whoever took her doesn't harm a single hair on her head. And as soon as we hang up, I'll be callin' Tiffany. I'll let you know what I hear."

I was only a couple of miles outside of Buxton city limits. I'd only have cell service for a short stretch of miles on the other side of town, so I needed to place this call while I could.

Thankfully, Tiffany had given us her cell number, and she answered the call instead of letting it go to voicemail.

"Deputy Roland," she said in a stiff voice. "To what do I owe the pleasure?" Only she didn't sound happy to hear from me.

Carly's meeting with her biological father hadn't gone well, so it made sense that Tiffany was pissed. She probably blamed Carly.

"Have you been in contact with Randall Blakely today?" I asked without preamble.

"Why are you asking?" Then she gasped. "What have you *done?*"

"*I* haven't done anything," I shot back. "But I need to know if you or that sorry excuse of a sperm donor told Blakely that you'd been in contact with Carly."

"I haven't. I swear. I haven't spoken to him since he called me after she ran from the wedding months ago."

"What about William?"

She hesitated. "I don't know, but he swore he wouldn't tell him."

"People lie all the time," I said. "A man who runs off and leaves his daughter with a monster stinks of cowardice to me. I have no doubt he'd sell her out, especially since he's disgusted that she's alive while her mother isn't."

"He said that?" she asked in horror.

"That and so much more, so I'm going to ask you again, do you think William Blakely would sell out his daughter?"

She was silent for several seconds, then said in a haughty tone, "I want to speak to Carly."

"So do I," I snapped back. "But someone kidnapped her about a half hour ago, and I'm trying to figure out who it could be."

"Are you saying she's in trouble with someone besides her father?" she asked. When I didn't answer, she cursed under her breath, then said, "I *knew* I should have been more insistent that she stay in Atlanta with me. I have the resources to protect her, while you obviously do *not*."

There was no point in arguing with her. She could have put armed security on Carly, but Carly had refused it.

Should I have tried harder to get her to stay?

Feeling guilty wasn't going to do shit to help her, though, so I shook it off.

"I need to know how likely it is that Randall took Carly, and you're the only person I know who can give me an answer."

She paused again. "I didn't contact him, nor has he tried to contact me," she said, less angry this time. "But William..." She drew a breath. "I need to make a call. I'll get back to you." Then she added, "Have you notified the police? I mean, outside of your own personal investigation."

"No," I said. "I'm trying to gather more information first. Besides, she's only been gone about a half hour. They won't take it seriously."

"Good. Don't. If Randall took her and he catches wind that you involved the authorities, there's a good chance he'll kill her and hide her body somewhere we'll never find it."

She had just spoken my greatest fear out loud. "I'll do *anything* to save her," I said, my voice breaking. "And if you want to convey that message to anyone who can help me get her back, feel free to do so."

She hung up without saying goodbye, not that I was surprised. I started to check Hank's phone to see if Carly was still on the highway or if her dot had disappeared, but the phone had locked up. I called the garage again from my own phone, and Hank answered this time.

"Chalmers."

"Tiffany swears she hasn't contacted or been contacted by Randall Blakely. She's going to talk to William Blakely to see if he squealed. She'll get back to me."

"That dick weevil probably called him this morning before he even drove out of the goddamned restaurant parking lot," Hank sneered.

I swallowed bile. "I'll let you know when I hear back from her, but in the meantime, I need the password to your phone. It locked up."

"Eleven-oh-six," he said, his voice scratchy.

I typed in the password, and the screen woke up. "That worked."

"Of course it did," he grunted.

"Have you found out anything on Wyatt's computer?"

"Nothing helpful. As soon as Max gets here, I'm having him take me out to my house to make sure there's no ransom demand on my phone. Then we'll follow your trail."

"I thought you wanted Max to stay in Drum to keep tabs on his family. He didn't sound keen on leaving."

"I changed my mind," he snapped. "And I can be pretty damn persuasive when I need to be. Is Carly still heading toward I-40?"

I checked the screen again. "It looks like they're about to get on it. Can't tell yet whether they'll take it to Asheville or Nashville."

"My money's on Nashville," Hank said. "Even though Wyatt's MIA, I'm starting to wonder if Blakely's not behind this after all. I'm not sure which of them to hope for. Wyatt's pissed as hell that Carly took his files, but from what I read in his notes about her, he still cares about her. And if it's her father…well, it's like I said before. If he didn't kill her outright, then he has a reason for keeping her alive."

It was the reason that scared me all the more.

CHAPTER FOUR

CARLY

We'd been driving for what felt like hours, but I had no semblance of time. When Wyatt had first abducted me, he'd driven in silence for a while, then turned on the radio. Based on the number of country songs he'd listened to, it had probably been an hour and a half. He'd duct-taped my mouth shut, and my hands and ankles were bound behind me with thin nylon rope, which he'd attached to a beam running along the side of the van. The setup forced me to face the back, presumably so I couldn't see where we were going. I snuck glances over my shoulder from time to time, not that it helped me figure out our location. Thankfully, the rope was loose enough that I could stretch my legs out. Once the initial terror of my situation had worn off, I closed my eyes and tried to focus on the turns he took and how far he was driving.

When he'd taken me, he'd said he was bringing me to my father, so if we were really heading toward Texas, he'd turn onto I-40. After a while he did, or at least I was pretty certain that's what happened based on his acceleration and the sounds of vehicles traveling fast.

Were Marco and Hank looking for me? Had they figured out who'd taken me? My best hope of being found by them sat in the front pocket of my pants. I had my new cell phone, and Wyatt hadn't figured it out yet. We rarely used phones in Drum, so he'd have no reason to search me for one. If I could get my hands free, I'd be able to call Marco or Hank and let them know that Wyatt had taken me.

My arms and shoulders had begun to ache soon after we left Marco's, and now the strain was becoming unbearable, and my legs and butt were sore and falling asleep from sitting on the hard floor. Did Wyatt plan to keep me tied up like this all the way to Texas?

Those weren't my only discomforts. I hadn't been to the bathroom in a while, and the pressure on my bladder was becoming intolerable. If Wyatt let me use a restroom somewhere, I could text Marco or Hank from behind the door. That gave me hope. Surely he'd stop somewhere soon. Still, it took another several songs before the van began to decelerate and then turned onto another road. We drove for a little while longer before Wyatt lowered the radio volume and turned off the road onto a bumpy surface. A dirt road? Looking over my shoulder, all I could see was the twilight sky.

My ears strained for any sounds from other people, but I only heard the distant sound of cars whizzing by on the interstate.

Where was he stopping? Did he plan to kill me and take my body to my father?

My heart began to race, but I convinced myself he wouldn't do that. If he'd wanted to kill me, he could have done it before now. Maybe he was stopping somewhere to spend the night.

The van came to a stop, and I heard him open the door. He got out and closed it. I expected him to open the back doors, but after ten seconds or so, I was still sitting alone in the van, the back doors closed.

Oh, God. Was this a meeting point he'd arranged with my father? Was I about to be handed off to his men?

I didn't hear any voices, only the cars, but it didn't stop my anxiety from ratcheting up. About a half minute later, the back doors opened, and dusky evening light filled the back of the van. Wyatt stood in front of me, glaring at me. A field sprawled out behind him, with no buildings or vehicles. No sign of human life.

Anger filled his eyes. "I had to pee, which means you probably have to pee too. But don't do anything stupid, Carly. I *will* shoot you. I've got nothing to lose."

I didn't make a sound from behind the tape or give any other kind of reaction, but my mind was racing. This could be my chance to contact Marco. Was there cell service out here?

He untied the rope that anchored me to the van, then grabbed my ankles and dragged me across the van floor so our faces were only a couple of feet apart. He searched my face as though looking for something, and I tried my best to keep my fear and anger hidden.

I refused to give him a reaction.

"I'm going to untie your ankles so you can pee next to the van, and after you're done, I'll untape your mouth so you can drink some water."

My eyes flew wide in shock. How was I supposed to pull down my pants, let alone squat and wipe, with my hands tied behind my back? Did he not plan on giving me any privacy? More importantly, this would make it impossible to text Marco.

I shook my head vigorously, my eyes wide.

His expression hardened. "I'm not taking any chances of you running. While I *will* shoot you, that doesn't mean I *want* to. I presume your father would prefer for me to bring you in alive."

I'd rather pee myself than let Wyatt take off my pants, but I couldn't tell him that, so when he tried to untie my ankles, I jerked them away.

He looped his hands around the rope and dragged me back. "If you have to pee, this is the only way it's going to happen."

I slowly shook my head.

He shoved me back inside and slammed the doors shut, but they opened an instant later. He dragged me closer and then ripped the tape off my mouth.

I hadn't been prepared, and the tape pulled tiny hairs from my face, bringing tears to my eyes.

"Okay," he said gruffly. "I'll untie you, but you have to swear to me that you won't run away. If you do, when I catch you—and we both know I'm a hell of a lot faster than you—I'll hogtie you and won't let you have water until we reach Texas." His eyes bored into mine. "Do we have an understanding?"

I still didn't want to give him the satisfaction of a reaction, so I said nothing.

"What was the point of taking the tape off your mouth if you're not even going to talk?" he demanded angrily.

I gave him a blank stare.

He released a string of obscenities, stomping away from the van, and then came back and started to untie my ankles. I didn't move, just let him slide me sideways and untie my hands.

"Okay," he said, when he pulled the rope free.

I slowly moved my arms in front of me, rolling my shoulders to try to loosen them up.

"You can pee," Wyatt said, "but fair warning, I'm not letting you out of my sight."

There was absolutely nothing sexy about peeing, so I decided I had nothing to lose. I slid off the edge of the van and nearly fell when my legs gave out. Wyatt grabbed my upper arm to keep me upright, and I released a cry of pain before I could stop myself.

His grip loosened, and I made myself walk forward a few feet. There were no trees in sight and the weeds were only a couple of feet tall. The only way I'd have any privacy was if he turned his back, which wasn't happening.

Which meant I couldn't use my phone.

I smothered my disappointment as I turned my back to him and unfastened my pants. My arms were still numb, so it took multiple efforts before I got them undone. I let my pants drop to my ankles, praying my phone didn't fall onto the ground. It didn't, but if Wyatt was watching closely, he could probably see one side of my pants were heavier than the other. Maybe he was too busy staring at my bare ass. I hated that he could see me like this, but I didn't waste any effort trying to cover myself. I squatted in the trampled weeds—my thigh and calf muscles screaming in protest—and kept my gaze on the field in the distance. Wyatt was right. If I took off running, he'd catch me before I found any signs of civilization.

When I finished, he surprised me by moving up slowly behind me and handing me a couple of tissues over my shoulder. I wiped myself and then tossed them into the field—feeling guilty as hell, but I had no intention of stuffing them into my pocket, and I was pretty sure Wyatt wouldn't throw them in a trash can. I pulled up my pants as I stood, careful to keep my cell phone out of sight, and fastened them with my back still to him.

"I know you're mad at me, Carly," he said, sounding exasperated.

Was he kidding? Of course I was mad at him. He'd killed Ruth and Franklin. He'd lied to me from Day One. He'd used me and kept secret notes on me. He'd kidnapped me at gunpoint, and none too gently at that. And then he'd tied me up and made me pee in front of him. Saying I was mad was the understatement of the decade. I wanted to turn around and scratch his eyes out, but I knew I didn't have any hope of getting away from him. My lack of anger was also making him uncomfortable, and he *deserved* to be uncomfortable. I turned to face him and stared at him with a blank expression, waiting.

He pushed out a sigh, then walked over to me and gently grabbed my upper arm, leading me back to the van. Once there, he grabbed my waist and hauled me up to sit on the edge.

Tears sprang to my eyes, remembering a different time. He used to be gentle to me. Kind. He'd made me believe he cared about me. In the beginning, I'd thought I'd found a man I could trust. But it had all been a lie.

A tear slipped down my cheek, and I looked up at him. "Was any of it real?"

He looked surprised. "What?"

"You made me believe you liked me, maybe even loved me, Wyatt. Was any of it real?"

Guilt filled his eyes. "Carly."

I choked back a sob and shook my head, turning my back to him and holding my arms together. I couldn't handle this right now.

He tied my hands and ankles, not cinching it as tight this time, while I continued to silently cry.

"I need to tape your mouth shut, but I don't want to do it while you're crying," he said softly.

"What difference does it make?" I asked, sounding defeated.

"You might not be able to breathe."

"Dead or alive, Wyatt," I whispered. "He won't care either way."

"But I do."

Hope bloomed in my heart. He hadn't completely become a calloused monster. The man who'd shown me tenderness was still in there somewhere.

I drew in a deep breath, trying to calm down. There was a chance I could convince him not to move forward with his plan.

"I only have a little bit of water left," he said. "Let me get it. When we stop for gas, I'll get some cold water bottles and food."

I didn't respond and kept my gaze on the floor of the van.

He walked away, leaving the back doors open. I heard the front door open and then close, and he reappeared holding a nearly empty water bottle. Leaning in, he held it up to my mouth. I tilted my head back to let the water slide in, but some dribbled down my neck.

Wyatt wiped the water off my chin, then lifted it so that our gazes met. "It didn't have to be this way," he whispered.

I dropped my gaze, overwhelmed with sadness that he'd resorted to these extreme measures. How had he sunk so low?

Picking up a roll of duct tape, he ripped off a piece and spread it over my mouth, pressing on the edges to make sure they stuck.

His voice hardened. "I'm going to stop in Knoxville. I'm going to leave you inside the van, but if you cause a ruckus and someone tries to help you, whatever happens to them will be on *your* head, Carly. Got it?"

So much for finding his heart.

He slammed the back doors, then got into the front. Soon we

were back on the highway, but he'd only listened to six songs on the radio before he stopped again. The sun had mostly set, making it darker outside. When he reached our destination, outside lights flooded the front of the van, leaving the back shrouded in darkness. I looked over my shoulder to see the top half of a travel gas station, but I was sitting low enough I couldn't see any cars, and I was sure no one could see me. It looked like we were under an awning, presumably next to a gas pump.

People were talking outside of the van, so close that I could probably touch them if there'd been a window in the side of the van.

"Remember," Wyatt said in a deep growl as he glanced back at me, catching me looking at him. "Keep quiet or you might get some people hurt. Ruth got in my way, and I killed her. I won't have any problem killing someone I don't know. Stay down so no one sees you and stay quiet."

The reminder that he'd killed Ruth filled my heart with agonizing grief, but I gave him the same blank stare as earlier.

He may have shown some tenderness out in that field, but he was a stone-cold killer. I couldn't forget that.

His mouth pinched with irritation. He looked like he was considering driving away, but he turned off the engine and got out. I didn't see him go into the store, so I presumed he was at the gas pump. I couldn't do anything with him standing there, but he'd said he was getting food and water too, which meant he'd leave me alone out here to do it. Once he was inside, I could scream and kick and pound on the side of the van. Of course, my screams would be muffled, but I'd make *some* kind of sound, not to mention the pounding was sure to get someone's attention. But was it worth risking someone else's life to save my own?

A few minutes later, he got back in and turned on the engine. When he pulled out of the space, I glanced over my shoulder at him in surprise. Wasn't he going inside? Then he pulled around to the back of the lot, in a space facing the highway. It was darker back here, and likely not close to anyone who could hear me. Tears stung my eyes as I stared at the back doors.

"Don't do anything stupid while I'm gone, Carly," he said, impassioned. "No one can hear you, and if someone does, I'll shoot them and drive away. And if I come back and hear you making any noise, I'll knock you out so you can't make trouble the next time we stop. Got it?"

Keeping my gaze on the back doors, I swallowed a sob and didn't respond.

I heard his impatient huff, then his voice rose. "*Got it?*"

Would he really do that? Punch me in the face and give me a concussion? Hit me over the head with a blunt object? He'd killed Ruth and Franklin, so why not? Still, while I was trying to hide my anger and fear, I couldn't bring myself to be subservient to him either.

"Goddammit, Carly," he shouted. "Do you fucking understand or not?"

I willed myself to stay perfectly still, clamping down every muscle in my body so it didn't shake with terror.

He released a grunt of frustration, then the driver's door opened and slammed shut. I steeled myself, prepared for him to fling open the back doors, but they stayed shut. I counted to thirty, and when he still didn't open the door, I decided he must have given up on getting a response out of me and gone inside.

Which meant I had less than five minutes to figure out a way to get free and run away.

Like before, the ends of the rope looped through the support bar on the side of the van were tied to my feet and my hands, but he'd left enough slack that I could try to loop my arms under my butt and bring my hands to the front. If I managed to untie the rope binding my feet, I could run away and deal with untying my hands later.

I started to shimmy, rocking to work my hands under my butt, but I realized that while the bindings weren't as tight as before, they still weren't loose enough for me to get my hands all the way under my butt. I pushed harder, straining the tendons enough to send sharp pain shooting through my shoulders. I stopped and took a breath. If I couldn't get my hands to my front, then I couldn't get free. Which meant I was stuck.

Tears burned my eyes again, but I drew in a deep breath through my nose and forced myself to calm down. Wyatt had been nice out in the field, probably because I hadn't fought back. If I continued to be stoic and non-defiant, then maybe he'd begin to trust me more. I might even be able to convince him that I was ready to face my father. If he let his guard down, then I'd have a better chance of finding an opportunity to break free and run.

But I didn't have much time. It was only a fifteen-hour drive from Drum, Tennessee to Dallas, Texas. Surely Wyatt would stop somewhere to spend the night and finish the drive tomorrow. Maybe I could find an opportunity then.

Had he already contacted my father, or did Wyatt plan on surprising him? If so, that might buy me an additional day to gain his trust, although I'd much rather break free from him in Tennessee than in Texas.

Several minutes later—long enough that if I'd had an opportunity to get free, I would have had time to escape—the driver's door opened, and the seat creaked as Wyatt got in and shut the door. I heard the rustle of a plastic bag and paper crinkling. The smell of a hamburger hit my nose, making my stomach growl.

"I got some food, but we're not going to eat yet," Wyatt said, sounding more in control. "There's a rest stop about an hour and a half ahead. The food will be cold, but it's the best I can do."

I didn't respond, just leaned my head against the side of the van and waited for him to get going. When we stopped, I planned to be the meek woman he'd always wanted me to be, but only so I could run.

CHAPTER FIVE

MARCO

Carly's dot on the map on Hank's phone showed her on 1-40 heading toward Nashville. If Wyatt had her, why would he be going west? Had Randall Blakely kidnapped her after all?

She'd started out about forty-five minutes ahead of me, but I'd brought that number down to thirty-nine minutes by the time I'd reached I-40. It didn't seem like much in the scheme of things, but they had to stop at some point. I'd catch up to them then.

Tiffany hadn't called me back. I'd tried to call her several times, but it had gone straight to voicemail. I'd figured it was a long shot she'd tell me anything helpful. Still, it was a disappointment.

I'd been on the road about an hour before I realized that Hank's phone was at less than 20%, and he used a different phone model than I did, which meant my charging cord wouldn't work. I needed to stop for gas soon anyway. Stopping meant losing the time I'd gained on them, but I assured myself that they would be stopping too, and Wyatt had no idea that I was so close.

I found a travel station on the east side of Knoxville and left the gas nozzle filling my tank while I ran in and peed, then found the charger. By the time I returned to the Explorer, Carly's dot had turned off the highway and was stopped somewhere off the road. I turned on the satellite view and saw they were parked in a field.

The blood rushed from my head, leaving me lightheaded. Had whoever taken her pulled off to kill her?

I started the car and peeled out of the parking lot, pulling back onto the highway and flooring the gas pedal. I drove close to ninety, weaving in and out of traffic until I got stuck behind two eighteen wheelers. I pounded my hand against the steering wheel in frustration. I was thirty-five minutes behind her. Would I be too late?

But then her dot started moving again. She got back on the highway, still heading toward Nashville.

I sucked in a deep breath of relief. Maybe her captor had let her go to the bathroom, although I couldn't stand the thought of someone holding her hostage and watching her pee. What was she going through? Was she terrified? Did she feel hopeless and alone?

But worrying wouldn't help anything. I continued to focus on her dot, moving steadily along I-40.

My cell phone rang, and to my surprise, it was Tiffany. I'd given up on hearing from her.

"Marco Roland," I answered.

"The good news is that I finally got William to take my call," she said without preamble.

"And?"

"After a moderate grilling, he admitted to calling Randall. But I don't believe he's the person who kidnapped her."

"Why not?"

"He says he didn't notify Randall until he got home about an hour ago. Wouldn't that have happened after she was taken?"

"Yeah." Which meant it was likely Wyatt after all.

"Plus, I don't think Randall knew where to look for her because William told him they'd met in Atlanta. Then Randall himself called me, demanding to know why I hadn't notified him that I'd spoken to her. He asked for her location, and I told him I didn't know any specifics. I only knew she was living somewhere in western Tennessee."

"But you know what county I work in," I said in surprise.

"I keep telling you that I want to protect her," she said, sounding pissed. "Why is that so hard for you to believe? I had to mix some truth in there so he might trust me if I needed more information from him. The only reason I talked to him at all was to find out what he knew about her, and from what I can gather, he was shocked to find out where she's been hiding. Of course, I'm sure he plans to send people to look for her."

"Yeah," I said absently. "I'm sure." So if Wyatt had abducted Carly, why was he headed west?

Then it hit me. He was taking her to Dallas.

"Did you get a sense of Blakely's mood regarding her?" I asked, my heart pounding against my ribcage. "Any chance he's had a change of heart?"

"The man is hard to read," Tiffany said, "but he can hold a grudge like nobody's business. So yes, I still think he plans to kill her."

"Thank you. If you hear anything else—"

"I'll let you know," she finished with plenty of snark. "And if you hear anything, you let me know as well. It's the least you can do." She paused, and when she spoke again, her voice was strained. "When I saw her yesterday…I was frustrated with her, and I let it show. If anything happens to her…"

"I'm going to find her," I ground out. "I'm going to bring her home."

"You won't be able to bring her back to Tennessee," she said. "Randall will be looking for her even more diligently now that he knows she's alive and has a general location. Plus, he knows she goes by Carly and that her hair is short and dark. William took a photo of her."

"What?"

"He admitted that he snapped it while we were walking in. I had no idea, and I definitely wouldn't have condoned it. But now that Randall has a photo, he'll be relentless looking for her."

"Then we need to find enough evidence to get him arrested."

"He's as slippery as an eel and buries his secrets. You'll have to dig deep to find anything."

"Then I'll rent a backhoe."

I hung up, then called Hank's satellite phone to tell him what Tiffany had said.

"I don't know anything new about Wyatt," he grumbled after I'd finished. "But I agree it's lookin' like he's the one who took her."

"He's heading west on I-40," I said, checking the dot on Hank's phone screen. "Approaching Knoxville. I'm about thirty minutes behind. I'm trying to gain on them, but I keep getting stuck in traffic."

"Don't do anything to get yourself killed in an accident," Hank said. "Carly's countin' on me to keep you safe."

"Since it's Wyatt who has her and not Randall Blakely, I think we should call the state police," I said. "They can be on the lookout for his truck."

"No point in that," Hank said. "We found his truck out at the state park."

"What?" I asked in dismay. "Where?"

"Near a trailhead. Looks like it was meant to be hidden, though. Not parked."

"So what's he driving?" I asked in dismay.

"Don't know. I know I said we'd be following you, but we're gonna head out to Bingham's to see if he might know."

"*You're* going to *Bingham*? What if he shoots you? He thinks you're about to re-take your old territory."

"It'll take a tougher man than Todd Bingham to get rid of me," he said with a dry chuckle. "And I know he respects Carly. He may not like her right now, but he respects her. Besides, Lula is bound to want to help her. That's why I'm showing up at his front door. Not for anything to do with his business."

"They're on the same property, Hank."

"I damn well know that, boy. But I'm takin' Max with me. Lula's not gonna let her newly discovered brother get shot. Which is why Wyatt might have gone to Bingham for a car."

"Doubtful," I said dryly. "There's no love lost between those two."

"Well, Junior swears he doesn't know about Wyatt having access to any other vehicles, so it's worth a shot."

"Speakin' of other vehicles, how'd you find Wyatt's truck?" I asked.

"I told you I've got people. They found it, and then Max and I checked it out." He paused. "Just finished searching it this second. We're close to Bingham, so I'm stoppin' there on the way out of town."

"How's Max doing with all of this?" I asked quietly.

"He's not talkin', and that boy's a talker if that tells ya anything."

"Yeah."

Max had been through hell the past few days, scratch that, the last few weeks. He considered his employees part of his family, maybe because his real family was so shitty. He and Ruth had been close, and I knew he thought of Carly like a sister. Top that with the arrest of his father and his mother taking off and leaving him to possibly face money laundering charges... But I was the only authority who knew about his involvement in the money laundering, and I had no intention of turning him in unless absolutely necessary. Other than Carly and me, the only person who knew Max's secret was his mother...and whoever had killed Ruth.

The sheriff's department didn't have any leads on her death, and neither did I. Did Bart Drummond have her blood on his hands?

If so, would he use the information about Max as leverage to get out of his own legal trouble? He'd had his own biological children murdered, so I had no doubt he'd turn on his son especially since Wyatt had always been the golden child in the Drummond household. But money laundering was a low-level crime compared to the murder charges Bart was facing. Juries tended to look down on the murderers of children, and the evidence against him was damning.

If he hadn't killed Ruth, someone else knew about the money laundering. Ruth had taken the spreadsheets from the restaurant home, planning to give them to Max's mother, yet they hadn't been in her possession when Max and Carly had found Ruth's body.

I glanced over at Hank's phone. Carly's dot was no longer moving. I picked it up and studied it for a few seconds longer. Not only had it stopped, but it was off the highway. I zoomed in and saw it was at a large travel center.

Had they made a detour to get gas?

Since I had plenty of fuel, I had a chance to close the distance between us. I found a break in the stream of bumper-to-bumper traffic and reached a stretch of road where I could safely drive at eighty-five. Carly's dot moved to a different part of the parking lot for about five minutes, then headed back onto the interstate.

I was now only eighteen minutes away.

"Hang on, Carly," I said out loud. "I'm coming."

CHAPTER SIX

HANK

I ended my call with Marco on my fancy new satellite phone and pushed out a sigh as I got out of Wyatt's truck. The dark mountain night seemed to swallow the noise with a greedy gulp, reminding me of my place in this world. I might think I was big and mighty, but someday soon, as was the way of things, I'd be worm food.

I was too damn old for this shit, but it wasn't like I had a choice. Someone had taken my girl, and I was tired of losing the people I loved.

So I squared my shoulders and ignored Carly's voice in my head saying my exhaustion was probably because my sugar was low. It was more likely because I was too damn old to be dealing with life and death crises. I'd thought I was giving them up years ago when I sold my business to Bingham, but I'd been a fool to think selling the business would mean I'd live a trouble-free life. Hell, it could be argued that I'd lost all my kin after selling. Maybe holding onto it had kept them safe.

Was I the reason so many of the people I loved had all met tragic ends? Maybe I really was cursed, like my ma had thought.

Back when I was a kid, my family had lived deep in the woods.

Our closest neighbor was homesteading on government property and making a meager living with homespun herbal remedies and love potions. My brother and I had stumbled upon her one day when she was out gathering her plants. He'd heard stories about her and gaped in fear. When she saw us, she threatened to curse us if we didn't run off, but I stood up straight, digging my bare toes into the mountain dirt, and told her she'd have to get through me to get to my brother.

So she'd cursed me.

My brother, of course, had run home and told our ma straight away. Ma, who took such things seriously, had sought out every home remedy known to destroy curses, but I hadn't taken it to heart. Even at the ripe age of seven, I hadn't believed in such things.

But as I got older and lost the people I loved, I'd started to wonder.

I shook my head. Why the fuck was I thinking about the past when I needed to be concentrating on the present?

A curse wasn't responsible for Carly being taken, but Wyatt Drummond was going to feel cursed when I caught up to him.

I grabbed the bottle of water Max had gotten for me and took a long swig.

Wiping my mouth with the back of my hand, I lowered the bottle and called out, "Max, come on out. We're goin' to pay a visit to Bingham."

With any luck, he'd be close enough to hear.

While I'd stayed back to search the truck, Max had taken to the woods to look for evidence that Wyatt had been out there. There'd been nothing in the truck to tell me where Wyatt was headed or what he'd been up to. He was definitely good at covering his tracks, the sonuvabitch.

"We're doin' *what*?" Max shouted from the trail. The sun hadn't set yet, but it was dark in the woods. I could see his flashlight beam bouncing in the trees.

"Wyatt ain't out there," I shouted. "He's headed west on I-40."

Several seconds later, Max appeared at the opening of the trailhead, staring at me with a look of disbelief. "How do you know that?"

"Because Randall Blakely only just found out that Carly was in eastern Tennessee about a hour ago. *After* Carly was taken. Which leaves your brother as the obvious culprit."

His face dropped, and my heart went out to the boy. While Max knew his brother had been up to no good, the truth still hurt. But he'd have to get over his hurt feelings and deal with the facts. We all did.

I chose to see it in our favor that Wyatt had taken her. Sure, he was heading toward Texas, which suggested he planned to hand her over to her father, but at least Randall Blakely didn't have her yet. Still, that didn't mean I trusted Wyatt to keep her safe from here to there. He was becoming unhinged and reckless, which made him dangerous and unpredictable, moving closer to the point of no return.

He'd already gone there as far as I was concerned, but Max, who had just realized every member of his family was a piece of shit, from his baby-murdering father to his manipulative, self-centered mother, who'd guilted him into dropping out of college and laundering money for her and then abandoned him to take the fall, was probably hoping at least one of his family members was redeemable.

He was shit out of luck if he thought Wyatt was a safe bet.

"Marco's following the signal on my phone," I continued, "but when he catches up to Carly, he won't know what kind of vehicle they're driving, which is why we're headed to Bingham's. We need to find out if he sold your brother a car."

He shook his head as he approached. "He'd never go to Bingham. They hate each other."

"Well, we're headed out there anyway. Bingham's been hoardin' information, and I aim to get some answers."

Max stopped in front of me and took a drink from the water bottle in his hand. "He's never gonna tell you anything."

"That's never stopped me from tryin' in the past."

Grimacing, Max said, "I'm scared to ask how you tried."

"You don't want to know," I grumped. "But I plan on talkin' to Bingham and hope he's forthcoming so I don't have to pull out my bag of tricks."

"You think you can torture Todd Bingham on his own property?" Max asked in dismay.

"I'm not rulin' anything out."

Swearing under his breath, Max got into the driver's side of his truck while I got in the other side. It took me two efforts to hoist myself onto the vinyl bench seat, but if Max noticed, he had the sense not to say anything.

"You know Bingham hates you," he grumbled as he started the engine.

I waved my hand in dismissal. "Don't you worry about that. Lula loves you, which makes you safe."

"Says you," he grumbled, pulling out of the parking lot. "I'm worried about getting caught in the crossfire."

Max headed out to Bingham's while I told him about my call with Marco.

He let out a low whistle. "So she's been outed with her dad, huh?"

I frowned. "Yeah. So we either need to bring the asshole down for good or she's gonna have to go on the run again."

Max's grip tightened on the steering wheel. "Then we take care of him. We brought down my father. We'll take on hers."

"I suspect her father's gonna be a lot harder."

He shrugged. "Whatever it takes." He shot me a glance. "You think Wyatt's takin' Carly to him for the reward?"

"Yeah," I said, feeling tired again. I was only in my late sixties, but I was feeling a hell of a lot older. "We have to count on Marco getting to her first."

And that stuck in my craw. I *despised* counting on other people to do my work for me. Sure, as a drug czar back in the day, I'd had to trust people beneath me to carry out tasks, but I'd always handled the big ones myself, and roping in Wyatt was huge. Nevertheless, the Wyatt situation was out of my hands, and I'd have to trust that the man who loved her would do everything in his power to save her.

I'd have to settle for confronting Bingham instead.

But once I was done with Bingham, I was gunning for Randall Blakely. After the hell he'd put my girl through, he was gonna face my wrath.

Right after I got my two hundred pounds of flesh from Wyatt Drummond.

Before checking out the report on the car, Max and I had stopped by my place to grab several weapons in case we came across Wyatt and needed them, but I knew better than to show up to the Bingham property hot. That was a sure way to start a war I wasn't interested in fighting. At least not yet.

Today was about Carly. I was counting on Bingham's goodwill toward her to get some answers.

Max must have been thinking the same thing because he gave me a sidelong glance. "Bingham may pretend otherwise, but he respects her. I've watched him come to see her at the tavern enough times to know. If we toss her name out right in the beginning, I think he might listen. And if not, we need to make sure Lula knows what's going on."

"Lula's not too happy with Carly right now. Not after their lunch." Carly had met Lula for lunch some days ago. Lula's mother had claimed to have evidence against Bart Drummond stowed away in a bank box, and she'd made Lula the signatory on the account. I had no idea how the woman had pulled that off since Lula was no more than eight years old when her mother was put away. Carly had hoped Lula would tell her what was in the box so she could use it to get Bart. But according to Lula, Bingham had taken whatever was in the box without showing her. All he'd said was that it was a bunch of nothing. When Carly had pressed for more information, Lula had gotten pissed.

Max frowned. "Lula may have been irritated, but she'll help Carly."

I grunted in response. I'd known plenty of women over the years and figured I could only trust half of them. Then again, I trusted fewer men.

Max slowed down as we approached the road that led to Bingham's property.

"You ever been out here?" Max asked, sneaking a glance at me.

Lord, this was bringing back plenty of unwelcome memories. I was tired just thinking about Bingham's father and the messes he'd stirred up. "It's been a while."

"Back when you were king, and Bingham was small fry?"

"I guess you could say that," I said, glancing out the side window. "The last time I was here was when we brokered a deal for Bingham to take over part of my operation."

Max's eyes flew wide. "No shit." I grimaced before he added, "He's worried, you know. He thinks you're gonna want to take over again now that you've got all your money back."

I shifted my weight on the seat. "Why the fuck would I want that headache? I'm content with my money. I don't need to be runnin' some big operation." I pushed out a breath. "Damn greed is the problem with this world. That and people trying to take what isn't theirs."

"One could say the latter is caused by the former," Max countered. "That they're actually one and the same."

"Maybe." I was no philosopher and I had no interest in twisting my brain, trying to think like one. Too much thinking and too little doing never got anyone anywhere.

I'd tried to be fair most of my life, but I was the first to admit I'd done plenty of things I had a right to be ashamed of. You couldn't run an illegal organization the size of mine without finding yourself on the slippery slope of moral ambiguity. Back in the day, I'd been sure I was making the right choices, but what had happened to my family was proof enough that I'd fucked up more than a time or two.

My wife, who'd stuck with me through all the business dealings, died from cancer. She'd found a lump but had waited to have it checked out because she knew I was stressed. My only child, dead from a drug overdose only a few years after that, an irony that had never been lost on me. Her son—my sixteen-year-old grandson—killed while trying to find his mother's drug suppliers.

A lump formed in my throat, a sure fucking sign I was getting soft.

There was no denying that I'd picked my business over family. Fate had laughed at me by taking them after I'd given my business up.

Maybe I *was* cursed.

After Seth was murdered in the motel parking lot last year, I'd thought my life was over. He'd been the last thing left in my life worth living for, and his life had been snuffed out, like a candle wick pinched between Bart Drummond's fingers.

We could never prove Drummond was behind it, but I knew the truth. And so did Carly.

Carly.

When she'd moved into my house after my leg was amputated, I'd gotten a helluva lot more than I'd bargained for. She'd brought sunshine into my life and given me a reason to live, and the thought of not seeing her bright smiling face every day ripped me to shreds.

No one could replace the family I'd lost, but Carly had wormed her way into my hardened heart, and now I considered her family too. If anything happened to her, I'd go scorched earth. I had absolutely nothing to lose. And I'd destroy anyone who got in my way.

Starting with Todd Bingham if need be.

Max slowed again as we approached the turnoff to the road that led to Bingham's property. "You got a plan?"

I steeled my back. "No. I'll figure it out as we go."

This wasn't something I could plan. The entire meeting would be reactionary. Bingham would likely be livid when I showed up at his front door. We'd always had an unspoken understanding that we left our personal lives out of our business lives. Not that he'd had much of a personal life before Lula and their baby. Now that he had something precious to protect, he'd come out armed and loaded.

Besides, I never ran my business meetings with a formal agenda. I knew what I needed to say and said it. That was my plan for this meeting too. Find out if Bingham had sold Wyatt a car. If not, did he know where Wyatt purchased it or what he might be driving?

Hell, this could be a mutually beneficial talk. If Bingham was agreeable enough, I'd consider telling him about Wyatt's secret files, including what was in the one on him.

Max grimaced but didn't say anything as he drove down the private road toward the structures on Bingham's property. The road curved, and around the bend was his house, an older bungalow. Further down the road, behind a locked, gated fence, were several large metal buildings that housed his auto salvage business.

"He might not be there," I said. "It might just be Lula."

"It's nearly dark," Max said, nodding his head toward the horizon. The sun had disappeared behind the trees but hadn't set yet. "Surely he'll be home."

I shrugged. You didn't get in a position like Bingham's without being a workaholic. I knew that firsthand.

Max pulled up in front of the house and turned off the engine. "I'll take it as a good sign that we haven't been shot yet."

I released a harumph and opened the door, getting my crutches in order and then sliding out of the truck.

I studied the house as I gathered my balance. The curtains in the front window fluttered, revealing a sliver of light, then closed, but the front door didn't open immediately afterward.

He was right. Lula was likely home alone.

"Max," I said, keeping my voice low. "Why don't you head on up to the door and let Lula know we're not here to cause a fuss."

He shot me a worried glance, then bounded up the steps and knocked on the door. "Lula. It's Max. We're here because Carly's in trouble and we need your and Bingham's help," he called out, his plea sounding desperate.

The door opened a crack a few moments later. Lula's face appeared, but her gaze was on me, not Max.

"I don't know nothing about 'er," she said hesitantly.

"We think Bingham might," Max said. "Is he here? We need to talk to him."

"I…" she said, her eyes wide with worry, only I didn't think her concern was about Carly.

A movement to my right caught my eye, and I saw Bingham coming toward us from the closest outbuilding, about seventy feet away. He closed the distance fast with his long strides. He carried a shotgun, and the tension in his body told me he was prepared to use it. "Whatcha doin' here, old man?" he demanded.

Putting all my weight on the tops of my crutches, I lifted my hands out to my sides. "I'm not armed. I'm no threat to you tonight, Todd Bingham."

"I can't think of a single solitary thing that would justify you bein' on my property," he snapped, looking like he aimed to kick my ass.

I continued holding my hands out at my sides as Max moved to the top of the porch. "He's not here to cause trouble," Max called out. "We're here because Carly's been kidnapped."

Lula gasped and slid out of the door and onto the porch. "*What?*"

"Go back in the house, Lula," Bingham shouted.

"The girl's not in danger from us," I said, holding his gaze. "We're here about Carly."

"I didn't take her," Bingham said, only about twenty feet from me now and showing no sign of slowing his pace. "So get back in your truck and get the fuck out of here."

"Todd!" Lula protested. "Carly's in trouble."

His jaw stiffened. "Not my problem. Now get in the damned house like I told you to."

"No," she said her voice hard. "Carly's helped me time and again. I owe her."

"The fuck you do," he said, coming to a halt about ten feet in front of me, holding the shotgun at my chest. His gaze was on me, not his wife. "Not my problem."

"It *is* your problem," Max said, walking down a step. "Wyatt Drummond was the one who took her."

Bingham's brow lifted ever-so-slightly, but otherwise, he looked unimpressed. "So? She got tired of the deputy and ran off with the bad boy." A sly grin spread across his face. "A tale as old as time."

"She didn't run off with him," Max said. "He kidnapped her, and now he's taking her to her father."

"Max," I grunted, pissed as hell that he was spilling a secret he had no business sharing, let alone with Todd Bingham.

Bingham tilted his head slightly. "Is that supposed to mean something to me?" he asked in a dry tone.

Max shot me an exasperated look. "What difference does it make if he knows? Her father already knows she was here. And Wyatt's taking her to him like she's a damn happy meal."

Bingham's brow wrinkled. "Is someone gonna tell me what the fuck is goin' on?"

Max's jaw tightened. "Carly is really Caroline Blakely, the missing oil heiress."

CHAPTER SEVEN

HANK

Bingham stared at Max for a long moment, then burst into laughter.

"It's true," Max insisted. "And my brother's taking her to her father."

I stayed quiet because he had a point. Bingham no longer posed a threat to Carly, but he might have information that could help her.

Bingham's laughter faded, and he turned his steely gaze on me. "Is this true, old man?"

"Yep."

Bingham's eyes widened slightly. "No shit."

"You know who Randall Blakely is?" Max asked.

Bingham snorted in disgust. "Hard not to. Not that I paid much attention to his daughter goin' missing, but the FBI's been mighty interested in him. How in the hell did she end up in Drum, Tennessee, and why is a rich girl like her working at your tavern, Max?"

"She was hiding," Max said. "Her father wants her dead. There's a reward on her head."

"If I recall correctly," Bingham said slowly, "there's a reward for her safe return."

"The optics wouldn't be so great if he said dead or alive," Max said in disgust.

Bingham tilted his head so he could study Max more. "Why's he want her dead?"

"That part's not important," I barked. "What's important is that she's in danger. We think Wyatt's on his way to Dallas to turn her in."

Bingham let the tip of his gun drop toward the ground. "Little Bart Jr. decided to give up on Daddy's money and get his own?"

"Looks like it," Max said bitterly.

"Carly's really rich?" Lula asked breathlessly. "She was always so broke. Her car was a piece of crap."

"*She's* not rich," Max said, turning toward her. "Randall Blakely never gave her money. She supported herself, but she couldn't bring much with her when she ran. So yeah, she's as broke as she appears to be. She hid in Arkansas for a while, but it looked like he'd found her, so she took off again. Her friends there got her a fake ID and a car. She was on her way to North Carolina when her car really broke down at the overlook. Wyatt towed her to his shop, and I gave her a job. I had no idea who she was. I only just found out a week or so ago, but Wyatt..." His voice broke. "Wyatt's known practically since she came to town."

"Fucking Wyatt Drummond," Bingham spat in disgust. "Why didn't he turn her in then?"

"Because he wanted her for himself," Max said.

"Why not turn her in when she hooked up with Marco Roland?" Bingham asked.

"I don't know," Max admitted. "I think he always had her identity in his back pocket, ready to pull it out when needed, but as long as she was in town, he wasn't in any hurry."

Bingham shot Max a dark look. "So why turn her in now?"

I grunted. "Because she found his dirty little secret."

"What dirty little secret?" Bingham demanded.

"We don't know his ultimate plan," Max said, "but he's been keeping track of everyone in town."

Bingham's jaw flexed, then he said through gritted teeth, "What does that mean?"

"He's got folders with notes on people," I said. "You included. Carly found 'em yesterday in a secret basement storage room in Wyatt's house. She stole most of 'em, and Wyatt figured it out. He was pissed as hell."

A storm of emotions swirled through Bingham's eyes, but he tamped them down. "I'm still strugglin' to see how this affects me."

"Todd!" Lula protested.

He ignored her, keeping his focus on me. "She's been a thorn in my side since Louise came back to town."

"Are you shittin' me?" Max snapped. "She got Lula the key to that safety deposit box!"

"Todd," Lula pleaded.

His face remained hard.

"Look," I said, not surprised by his attitude, yet pissed anyway. "I don't expect you to lead the charge in findin' her. What I want is information. Did you sell Wyatt a vehicle?"

Bingham barked out a laugh. "Drummond would never come to me for a vehicle. Heard he lost his shit when he found out I sold one to Carly."

I pressed my lips together. I had no reason to think he was lying. There were plenty of other places to get a vehicle, but I'd still hoped to get lucky.

"What do you know about Franklin Tate?" I asked, wondering if there was a connection in all this.

Bingham shrugged. "I know his wife is dead and that he's missin'."

Lula sniffed and held her hand to her lips as if holding back a sob. Carly had said they didn't get along much, but they'd still worked side-by-side for long enough for it to be a shock.

"Rumor has it you two had a beef," I said. I didn't know any such thing, but Carly had told me that Franklin had been hesitant to go to Bingham for help replacing his truck, so it stood to reason.

Bingham's mouth pressed into a thin line, then he said, "Don't know how you came by that information. He might have had a beef with me, hell, half the town does, but I didn't give the man a second thought." Then he added, "Didn't kill his wife either. And if you don't believe me, I challenge you to come up with a reason why I would."

He had a point. I couldn't see him going to such lengths to get proof Max and his mother were laundering money.

"Why are you researching my family history?" Max asked.

Bingham released a short laugh. "Goin' for broke, huh?"

"I'm sick of pretending shit's not going on around me," Max spat out. "My brother claimed to have been trying to bring our father down, but he was doing God knows what while spying on half the town. Now tell me why you're researching my family."

Bingham shifted his weight. "Seems to me *you're* the one spying, claiming to know what I'm lookin' into."

"He heard it from Carly," Lula said. "Carnita told her."

"Fuckin' librarian," Bingham grunted.

"Why are you researching my family?" Max repeated.

"Tell him," Lula said.

Bingham's attention shifted to his wife, and he growled, "Stay out of this."

Max stood taller and puffed out his chest. "Don't talk to her like that."

"What?" Bingham said with a laugh. "Now that you know you share some DNA, you supposedly care about her?"

"I cared about her before that," Max sneered, "all while *you* were using her and throwin' her out like trash, so shut the fuck up."

I lifted a crutch and pointed it at Max, then Bingham. "Everyone just calm the fuck down."

Bingham's gaze turned murderous. "Don't you *dare* tell me to calm down on my own property."

"We're losing focus," I said in a sharp tone. "Wyatt Drummond took Carly and is bringing her to her father. That's what's important."

"And I told you, I don't give a fuck," Bingham said.

"I do," Lula said, taking a step forward. Her arms were wrapped tight around her chest. "I give a fuck." She turned to face me. "What do you need?"

I let myself feel a sliver of hope. "I need to know what Wyatt's ultimately up to. What he planned to do with all those files."

She shook her head. "We don't know that Hank, I swear. We didn't even know about them."

Bingham's eyes burned with fury. "I'm not having any part of this. I'll give you five minutes with my wife, then you better get the fuck off my land." He pointed his gun at us to underscore the threat, then turned and stalked back to his garage.

Lula watched him walk away and picked at her bottom lip with her thumb and index finger.

"You're scared of him," Max said gently.

She looked over at him and gave him a tight smile. "Of course not. Don't be silly. He's just under a lot of stress."

"I'm pretty sure that's what most abused women say," Max countered.

She gave him a long look, then took a deep breath and stood up straighter. "We're wastin' time. He means it about the five minutes, so we probably only have about four minutes left."

Max's face twisted in grief. "Lula—"

Her eyes hardened, and she lifted her chin, turning all business. "Todd's been looking into property boundaries. He's not just lookin' in the library, but the courthouse too. He thinks the Drummonds took part of his family's land—the section where Bart's building the resort—and he aims to prove it."

"Is he deluded enough to think he's got a chance at getting it back?" Max asked incredulously. "They already went to court over this, and he lost."

"Todd thinks he can change their minds."

"He's dreamin', and he knows it," I said with a frown. Bingham was a lot of things, but stupid wasn't one of them. "What else is he up to?"

Her eyes widened as she stared at me. "What are you talkin' about?"

"He's not gonna hurt you," Max coaxed. "You can tell us."

She clasped her hands together and pressed them to her chest. "If he finds out I told..."

"It's okay," Max soothed, stroking her upper arm. "Tell us."

She nodded in my direction. "Hank's right. Todd's been lookin' into something else." She swallowed hard. "Hereditary stuff."

Max shook his head in confusion. "What kind of hereditary stuff?"

"His brother—Rodney—he had Rh-negative blood."

Max shook his head again. "What does that have to do with anything?"

"Rodney had surgery when he was really little and he had to have a blood transfusion. There was a shortage, and Rodney's blood type was rare. O-negative. Todd offered to give his brother his blood, but when they tested him, Todd turned out to be A-positive. Todd and his momma convinced his dad to get tested, and he was A-positive too."

Max narrowed his eyes. "So?"

"Rodney's momma was O-negative. There's a chance Floyd Bingham wasn't Rodney's dad. The negative part is recessive, so both parents need to carry it."

Confusion clouded Max's eyes. "I still don't get what that has to do with anything."

She shot an impatient look toward the metal building, then back at Max. "Todd is convinced that Bart is Rodney's daddy, and he's also sure the Rh-negative came from him."

"My dad's Rh-positive," Max protested, still looking like he had whiplash. "So are Wyatt and me. We're A-positive. We had our blood tested when we donated to a blood drive. Dad too."

"I'm negative, Max," she said, her eyes brimming with tears. "And Bart's definitely my father, so he must have the recessive gene. Todd was looking for proof that some of the people in your family tree were Rh-negative."

"How in the world would he know that?" Max asked in disbelief. "It's not like people knew their blood type a hundred years ago. And even if they did, they didn't post them in library books."

"No," she said. "They didn't, but when an Rh-negative woman has an Rh-positive baby, if her next baby is negative, the baby runs the risk of dying. There's something in the momma's blood from the first baby that attacks the Rh-negative baby. Beatrice is Rh-positive, so if I get pregnant again, I'll have to get shots to keep the antibodies in my blood from hurtin' the baby. But a hundred years ago, pregnant mothers had no idea they were negative, and even if they did, they couldn't get shots. So the babies died. Todd's been looking at babies' deaths in the Drummond line."

"Tryin' to establish a pattern," I murmured.

She nodded. "And he found one. But turns out he didn't need the research after all because he found proof."

"How?" I asked skeptically.

"DNA. Right after Bea was born, he was up in the attic, looking for some of Rodney's stuffed animals. He found an old box packed with all of Rodney's old clothes, but he also found a hairbrush. He realized it was Rodney's, and he sent it to a lab for genetic testing. They swabbed my cheek too, so they could compare us." She held Max's gaze. "Rodney and I matched as half-siblings. For the longest time, Todd thought his own father had killed Rodney and buried him somewhere. Now he's sure Bart did it."

Max rushed to the edge of the porch and vomited over the railing onto the bushes below.

Lula put her hand on his back and silently rubbed up and down until he stood back up and wiped his mouth, his color off.

"Is Bingham wanting to put Drummond away for murdering his brother?" I asked.

"He's got a lot of things in motion, but yeah, that's one of them," she said. "I know he found something in that safety deposit box. I think it's proof he killed another of his kids, but he won't tell me."

Max looked like he was about to be sick again.

"We already found proof that he killed a baby," I said, hit by a wave of exhaustion, this one stronger than before. I ignored it. "Ruth's mother had a baby, and Bart admitted on a video that he killed her."

Her eyes widened in horror. "Ruth?" she gasped. "Did Bart kill her?"

"We don't know who killed her," I said, "but Marco turned the video in to the state police and Bart's sitting in a jail cell in Knoxville on first-degree murder charges."

Her face paled. "What kind of monster would kill a baby?"

"He tried to kill *you*, Lula," Max said, resting his forearms on the porch railing. His voice sounded dead. "And if he killed Rodney, why wouldn't he kill a baby?"

"The point is," I said firmly, "we found the proof to put the bastard away. Proving Drummond killed Bingham's brother is a moot point." Still, I wasn't so sure I believed that myself. Drummond was as slippery as they came. A backup plan wasn't a bad idea. "Is Bingham still chasin' this down?"

"He's trying to figure out where he might have buried Rodney," she said. "He hasn't had much luck."

"I might know," Max said, his voice breaking.

Lula gasped. "What?"

Drawing in a deep breath, he stood upright. "When I was little—after Rodney disappeared—my mother took flowers to a clearing in the woods behind our house. She did it every week or so for about a year. I thought it was weird, and when I asked her about it, she said she was making a gift to the fairies." He scrubbed his face again. "Honestly, I forgot all about it until just now. She used to do all kinds of weird things, so it was just one of many." He turned to look at Lula. "I think maybe Rodney's buried out there."

Her face turned pale white. "I have to tell Todd."

Max was already heading down the porch steps. "Have him meet me at my parents' house with a shovel."

"Now hold on," I said gruffly, lifting a hand. "That's all well and good, but we still don't have anything to help us find Carly."

"I don't have anything that will help you, Hank," Lula said. "If I did, I'd tell you."

"But Bingham's been collecting information on Wyatt, just like Wyatt's been collecting information on him. Right?"

She didn't deny it. "I have no idea what Wyatt's up to, Hank. I swear on my baby's life."

Dammit. She was telling the truth. I could tell from the look in her eyes.

"Does Bingham have a theory about Wyatt's ultimate goal?" I asked.

"To step into Bart's shoes," she said. "Take over it all. But I have no idea how he's aimin' to do it. Todd never tells me that stuff. There's only one other thing…" she added, sounding unsure.

Max and I gave her our full attention.

"Todd thinks Wyatt knows about Rodney. He thinks he might have even helped Bart cover it up."

"Wyatt would have been a boy," I said. "Not much older than Rodney. Probably middle-school aged."

Lula twisted her hands in front of her. "He still blames him."

"So Bingham *does* have dirt on Wyatt?" I prodded.

"Like I said, I don't know any specifics, but I've heard Todd say Wyatt needs money. If he's turning Carly in to her father, then I guess he found a way to get it," she said bitterly. "Todd thinks Wyatt has some big plan in motion, but I don't know anything about it. I swear."

I nodded. Good enough for me.

I took off toward the oversized garage. "Bingham," I shouted. "I've got a proposition for you."

Bingham popped out of the building, obviously watching us from the shadows of the building. He held his hands out at his sides, holding the butt of his gun in his right hand as he headed toward me. An amused grin spread across his face. "Not interested. I'm a happily married man."

"A trade," I called out. "The body of your dead brother for everything you know about Wyatt's schemes."

His smile fell, and his body morphed into fighting mode, his shoulders stiffening. "You've known where my brother's been buried all along?"

"No," I said. "I had no idea what happened to the boy, but Max remembers his mother acting weird around the time the boy disappeared, so we might know where he's buried."

Bingham shot a glance past me at Max. "You part of his murder too?" he shouted.

Max released a bitter laugh. "No. My father considered me too soft. A pussy. He never told me a thing. My brother, on the other hand..."

Bingham's mouth pressed into a thin line.

"You want his body. I want information that'll help me help Carly," I said, my voice hard. "Let's work together to get what we both want."

"If you can produce my brother's body, I'll tell you everything I know about Wyatt Drummond, old man," Bingham said with a sneer. "But it's an all-or-nothing deal."

"Fine by me," I said. "Time's a wastin'. Get a crew together and meet us out at the Drummond property in an hour. Bring a bunch of shovels and lights. The sooner we get started, the sooner I get my girl back."

Bingham eyed me suspiciously, but after a moment he said, "You better not be screwing me over, old man."

"Funny," I grunted, "I was just about to say the same to you."

CHAPTER EIGHT

CARLY

We drove for about another hour before Wyatt stopped again. He exited the highway and drove for less than a minute before turning again, backing up, and coming to a stop.

I looked over my shoulder to figure out where we were, but all I saw was darkness. The sun had set, and there didn't seem to be any streetlights.

"I think it's safe to eat here," Wyatt said as he turned off the engine. He opened his door and got out. Seconds later, the two back doors opened.

I blinked as my eyes adjusted to the view. The back of the van was facing brush and trees. I could hear the whizzing of the cars on the interstate to my left, closer than when we'd stopped in the field.

Without saying a word, he placed two bags on the van floor, one paper and the other plastic. Something in the plastic bag thunked dully when it hit the metal. Pulling his phone out of his pocket, he turned on the flashlight, then propped it against the inside wall of the van, providing some illumination. He stood in front of me and stared into my face for a few moments before he lifted his hand and gently pulled the tape from my mouth, tossing it to the ground. "I'm gonna untie your hands so you can eat, but not your feet. When you're done, I'll let you pee."

I took in a deep breath through my mouth, feeling like I could breathe freely for the first time since he took the tape off before. My mouth was unbearably dry.

Turning slightly to the side, I lifted my hands about an inch, ignoring the pain shooting through my shoulders.

He moved the flashlight closer and got to work untying the knots securing my hands. "We won't go much farther tonight," he said, his fingers brushing against my hands and wrists as he worked. "I just wanted to get some distance between us and Drum."

"Do you really plan to take me to my father?" I asked quietly, my voice rough.

He continued working on the knots, ignoring my question.

"He'll kill me, Wyatt," I pleaded. "Is it money you want? Or do you want to be part of his criminal organization? Because I'm pretty sure that's over."

"My reasons are none of your concern," he said insistently but without any anger. He pulled the rope off my wrist and tossed it to the side.

I slid across the floor so my still bound legs hung over the edge of the van and searched his face. "Forgive me if I think getting murdered *is* my business." I held back a retort that Ruth had probably thought it was her business too. Yes, I was pissed as hell, but anger wasn't going to get me out of this. Fake subservience just might.

"I don't think he'll kill you, Carly," he grunted.

"He killed my mother and her parents," I said, "and I was working on proving it."

"It doesn't matter," he said, sounding resigned. He pulled a water bottle out of the bag and twisted off the cap.

What the hell had happened to this man? Did he actually believe my father wouldn't kill me, or had he just convinced himself of that so he could live with his decision? Had he come to some similar reasoning before killing Ruth?

Still, I wasn't convinced killing Ruth had been easy for him. There was no denying the haunted look in his eyes. Ruth and Franklin's murders hadn't come without a price.

The Wyatt I'd met last November wasn't a cold-blooded killer. Despite all his secrecy and lies, I still believed he wanted to protect me, and the notes in my file backed that up. I needed to bring out his protective nature, which meant continuing to act like a helpless fool.

"Can I have a drink of that water?" My mouth *was* unbearably dry, but I needed him to think he had to take care of me.

He held the bottle toward me. I took it and swallowed in greedy gulps. My shoulders ached, making it difficult to lift the bottle high enough. My arms shook and water dribbled down my chin and neck.

"Carly, let me help," he said softly, taking the bottle and lifting it to my mouth.

I nearly stopped him, but I needed him to soften toward me, and this might help, even if this felt infantilizing

I continued to drink; half the bottle was gone before he pulled it away.

"If you drink too much, you might get sick."

"My mouth's dry from that tape," I said.

"Well, we can't be stopping to pee every hour," he said, irritation creeping into his voice, "so that's enough for now." He screwed the cap back on, then pulled two burgers out of the greasy paper bag, handing one to me. "They're sure to be cold by now, but it's probably not the worst thing you've eaten." His mouth lifted with a hint of a grin. "You've eaten enough at Max's."

I wasn't sure how to respond. I wanted to tell him to keep his brother's name out of his mouth, but instead, I ground my teeth and slowly unwrapped the burger and then leaned over to take a bite.

"Are your shoulders really that sore?" he asked skeptically.

"You try being hog-tied for a few hours, and then I'll come back and ask you that question."

Shit. *Rein in the hostility, girl. Save it for when you find a way to drop-kick his ass.*

"You're not hog-tied, Carly," he said in resignation. "I can show you hog-tied if you'd like." Thankfully, there wasn't any heat behind his words.

"Wyatt, he really is going to kill me," I said softly after I swallowed a bite. "*Please.*"

He leaned against the back edge of the van and took a bite of his burger, staring at the trees.

"You know I want to bring my father to justice," I said. "Let's do it together."

He glanced over his shoulder. "You said that when you came to town last November, but from what I've seen, you've spent the majority of your time focusing on *my* father."

"And in the beginning, you told me that was what you wanted, Wyatt. You said you wanted to bring Seth's killer to justice. We agreed to take down your father and then go after mine."

"Carson Purdy was Seth's killer," he said insistently. "And he met his justice at the end of a gun."

"You know your father was behind it. Him and his greed." Talking about this was pissing me off all over again. "You wanted him gone too!"

"I wanted him gone in my own time," he said, his voice tight. "But you just kept gettin' in the way."

"So what was your plan?" I asked, blowing my own intention to remain calm. A year ago, it would have been only too natural to play the good, dutiful girl who wouldn't make waves. That was who Caroline Blakely had been. A doormat. But going on the run had changed me, and that woman was gone. Carly Moore couldn't find it in herself to play the subservient fool for long. I was done kissing up to men.

"Do you plan to take over your father's empire?" I demanded.

"Because from what I've heard, there really isn't much of an empire left. The only thing he has of worth is the resort, his one last hurrah."

"That resort is a good idea," Wyatt snapped, then took a bite of his food.

"Yes," I admitted. "It is. It can bring a lot of jobs to the area, but where do you come into all of that? What was your plan, Wyatt? Why were you taking notes and keeping files on everyone in town?"

"Not everyone," he grumbled.

"Just the people you saw as threats?" I suggested. "As many files as there were in your basement, it must be exhausting living with that much paranoia."

He groaned and took another bite. Then he abruptly turned and reached into the bag and pulled out a carton of fries and put them on the floor next to me. "I forgot your fries."

"You don't have to *do* this, Wyatt," I said, returning to pleading. "Tell me what you're looking for from my father, and maybe I can help you get it without getting either one of us killed. Because I can't believe you'd give me water and a burger and fries if you didn't care whether he killed me."

He took another bite, keeping his gaze on the trees.

"If you let me go, I won't tell Hank or Marco that you kidnapped me. I won't tell them anything. And I'll do everything in my power to get you what you want."

"Then what exactly will they think happened to you?" he scoffed, shaking his head.

"I'll tell them I went with you willingly."

He snorted. "They'll never believe that."

"They might buy it," I said. "They know I want you to give me answers. They wouldn't like it, but I can make them believe it."

He shook his head. "You have your answers. You took my files. Besides, I already told you that I killed Ruth." His voice broke, then he drew in a deep breath, setting his burger down on the van floor.

"What really happened with Ruth? I'm having a hard time believing you killed her in cold blood."

"Because you don't want to believe I can kill you too?" he asked, his icy tone sending a shiver down my spine.

"No, because I know you *did* care about her, and I still believe there's a decent person in you. You've just covered it up with hate and the need for revenge."

"You believe there's still a good person inside of me?" he scoffed.

I reached over and grabbed a handful of his T-shirt and tugged.

He turned to look at me.

"I believe that, Wyatt." But I couldn't help thinking that I only *wanted* to believe it. He'd gotten Seth killed by encouraging him to spy on Carson Purdy's drug ring. Still, he'd been genuinely upset by his death. He'd paid for his funeral. He'd looked after Hank. There was no faking that.

And, yes, there was also a part of me that simply couldn't accept that I was sitting here eating cold burgers and fries with a hard-hearted murderer.

So how did I spin my kidnapping to fit the version of Wyatt I wanted to believe? How did I justify all the deceptions and outright lies?

My stomach churned, and I set my burger down.

"You have to eat, Carly," he said, irritation creeping into his voice again.

I pushed out a sigh. "What difference does it make?"

He squatted in front of me, cupping the side of my face with his hand. "We could have been good together," he said softly.

I released a bitter laugh. "Until I pissed you off, and you took me to my father for your perceived fat reward."

His eyes hardened. "There's nothin' perceived about it. There's a reward waiting for me in Dallas."

"He's never gonna pay you," I said in disgust. "He's a hell of a lot more likely to kill you."

"I don't believe you. You'll say anything to keep me from doing this."

"Why are you interested in collecting that reward now? Why not turn me in back when you first realized who I am?"

He didn't answer, just stared into my eyes, searching for something. What did he hope to find? Love? Devotion? Regret?

I still believed he cared about me, but I was too emotionally exhausted to deal with it right now. I needed to regroup and come up with a plan. But I also couldn't wait too long because depending on the length of time we stopped later, we could be at my father's front door by tomorrow night. And I couldn't do much persuading while my mouth was taped shut.

"I'm tired, Wyatt. It's been a hell of a day."

"I'll bet," he said sarcastically. "What with stealing my files and then getting my father arrested."

"I met my birth father this morning."

His eyes widened in surprise. "What?"

I told him about William's relationship with my mother and what we'd learned about my grandparents' death.

"That could have been coincidental," he said.

I gave him a dry look. "And it's coincidental that their accident was nearly identical to my mother's?"

He didn't respond.

"William hates that I lived and my mother died."

"Sounds like an asshole."

"Hank punched him."

He laughed. "I would have liked to have seen that."

"I wish I'd recorded it," I said dryly, then took a breath, preparing to start in on him again. "You're wrong about my father not wanting me dead. He already tried to kill me, Wyatt. I was supposed to be in the car with my mother, but at the last minute, she left me at home. He'd planned for us both to be in that car."

He glanced back out at the trees. "That was a long time ago."

I released a bitter laugh. "Guess our fathers have a few things in common. They both have no qualms about killing their children."

"Carly."

"What happened with Ruth?" I asked, careful to keep my anger out of my voice.

For a moment, I thought he was going to answer, but then he stood, turning to put his trash in the paper bag. "Are you really done with your food?"

I shoved the wrapper toward him. "Yeah. I don't have much of an appetite." Instead, I reached over, grabbed the water bottle and took a long drink. Thankfully, my arms had started to loosen up.

He opened his mouth to say something, then stopped, grumbling something about "stubborn" under his breath. "You need to pee before we go."

My feet were still hanging over the edge, and to my surprise, he lifted them into the van floor and started to untie the rope wrapped around them. He tossed the binding next to the food bag, then took off my shoes and set them on the floor.

Why was he taking off my shoes when I was about to get out and pee?

Grabbing a couple of napkins, he handed them to me. "You still have to pee in front of me, but I'll turn to the side to give you a bit of privacy."

I started to slide off the edge of the van. The second my feet hit the ground, I realized why he'd taken off my shoes and said he'd give me privacy. The ground was covered in gravel. If I tried to run barefoot, Wyatt could easily catch me even if I darted into the brush ahead.

My legs ached, and I took a tentative step, waiting for the stiffness to subside.

Wyatt reached over and grabbed my waist with both hands to steady me, leaving me staring at his chest.

"I need you to trust me, Carly," he whispered. "Can you do that?"

Part of me wanted to. When I first came to Drum, he'd taught me to hope when I'd forgotten how, but that was a long time ago. This was the man who'd murdered Ruth, whatever the circumstances. The man who'd kept a secret basement room with shackles embedded in the floor.

Wyatt was not to be trusted.

Still, I had to make him believe I was falling for his bullshit.

"Yeah," I whispered back, keeping my gaze on his chest. "I'll try."

He pushed out a deep sigh, and for a moment I thought he was going to hug me, but he took a step back and dropped his hands to his sides.

My heart racing, I took several steps closer to the edge of the trees, gravel chunks digging into the bottoms of my feet. Each step felt like it brought me closer to potential freedom, but it could also put me in mortal danger if Wyatt realized I was even thinking about escaping.

Could I make this happen?

The rocks hurt my feet, but I could still try to run into the woods. I had no idea how deep they went, but it was dark, and I had the possibility of hiding. However, as I stopped at the edge and started to unfasten my pants with shaky hands, I realized the ground was covered in broken twigs and stiff grass. My feet would likely be cut to shreds, and even if I could ignore the pain, I'd probably leave a bloody trail for Wyatt to follow.

I had to find another way.

True to his word, Wyatt turned sideways and stared out toward the highway, which meant he might not notice if I dug out my cell phone and texted Marco and Hank. They might not get the text immediately, but at least they'd know who'd taken me and where we were going.

With my button and zipper undone, I slid my pants and panties to my ankles. Squatting, I slipped the phone out of my pocket and hid it between my lower legs as I typed in the password to unlock the screen, glancing back and forth between Wyatt and my phone.

When it opened, it cast a bright light, and I scrambled to dim the screen.

"Why aren't you peeing yet?" Wyatt asked, still facing away.

"Performance anxiety," I said, my pulse pounding in my head. "It's hard to pee when someone's listening for it." I pulled up my text screen. I'd only had my phone for a day, but thankfully, it was nearly identical to the phone I'd had before my life fell apart. I pulled up the texting app and inserted Marco and Hank's names, which I'd thankfully already added to my contacts. I started to type as I peed, trying not to fall over.

Wyatt kidnapped me. He's taking me to my father.
He killed Ruth

I looked over at Wyatt and pressed send just as I stopped peeing. Then I typed.

Hour west of Knoxville on I-40.

Wyatt glanced over and did a double take when he saw the light reflecting on my lower leg. He hesitated a second, then bolted toward me.

I jerked upright, pulling up my pants, but I dropped my phone in the process.

Wyatt leaned over to reach for it, but I kicked him hard in the side with the heel of my foot, throwing him off balance.

Shit. What was I going to do now? I hadn't planned on trying to escape yet, but I also didn't want to stay and deal with the fallout of pissing him off.

Wyatt fell over to the ground, and I took off running toward the interstate. We were only about a hundred yards to the entrance and exit ramps, and even if I couldn't make it to the highway, maybe someone would exit the highway and see me.

It was a long shot, but it was all I had.

My pace was slowed down by the rocks digging into my bare

feet, but I made it to the road before I looked back to see Wyatt getting to his feet with my phone in his hand. Obviously, he'd taken a second to read my texts. There was no one out here, so he must have figured he could spare a couple of seconds before he gave chase.

The asphalt was smoother, so I dug in, putting all my energy into racing down the middle of the road toward the entrance ramp. I could hear Wyatt's grunts of frustration growing closer behind me. I was sure he would catch me any second, but then I saw an SUV pulling off the highway onto the exit ramp.

I started waving my arms, shouting, "Help me! *Help!*" Praying that Wyatt didn't do anything stupid to try to hurt them.

"Goddammit!" he shouted, sounding like he was nearly an arm's length away.

The SUV sped up. When I could make out the person behind the wheel, I was shocked to my core.

Marco?

I nearly wept, torn between believing I was hallucinating or that I was about to be saved. How had he found me so quickly?

Marco slammed on his brakes. His tires screeched across the pavement as he skidded to a halt in the middle of the road.

CHAPTER NINE

MARCO

I couldn't believe my luck when Carly's dot pulled off the highway and stopped at a location right off the exit. It didn't look like a gas station, but I didn't care what was out there. I only hoped they'd stick around long enough for me to intercept them. I was only about ten minutes away. Still, I had to expect trouble. Wyatt wasn't just going to let me swoop her up and into my car. And on the off chance her father was responsible, I might find myself facing a whole team, not just one man.

My lame plan was to show up and try to assess the situation, then call the state police if necessary. At the very least, I could get the license number of the vehicle that had taken her. That would be something.

I couldn't believe my luck. The dot didn't move.

I was about a half a mile from the exit when a text popped up on Hank's phone.

> Wyatt kidnapped me. He's taking me to my father.
> He killed Ruth

Wyatt? That definitely made this a lot easier to deal with.

I floored the gas pedal, driving onto the shoulder to get around a slower car in front of me, then wove back onto the highway.

The exit ramp was in sight, about two hundred feet ahead when another text came through.

Hour west of Knoxville on I-40.

That was my girl, giving me clues to her location. Little did she know I was closer than she thought.

I barreled toward the ramp, then started to slow down. If I showed up driving like a maniac, he'd immediately realize who I was and why I was here. But if I drove past and assessed the situation, I could come up with a plan.

I'd just started up the exit ramp when I saw Carly running toward me. A man followed close behind her, only a few arm lengths away. Wyatt.

My blood boiled with rage.

Change of plans.

I was going to kill the fucker.

I gunned the SUV and hit the window button to roll it down. Grabbing the gun out of my holster, I held it with my right hand as I slammed the brakes, hanging onto the steering wheel with my left hand as the car fishtailed. Carly came to a halt, bringing her hands up to her head as the car came to a stop about ten feet away from her.

Wyatt ignored me, closing the distance between them.

"Don't you fucking touch her, Wyatt!" I growled, throwing my car into park, and opening the door.

Wyatt looked just as shocked to see me as Carly had been, but it didn't slow his reaction speed much. He grabbed Carly by the arm and jerked her in front of himself, then wrapped an arm around her waist to pull her back to his chest, holding a handgun to her temple.

"Careful, Marco," he grunted, huffing to catch his breath. "You don't want to shoot her."

I pointed the gun at his head. "*I* don't intend to, but it appears I can't say the same thing about you."

Wyatt started to laugh.

Carly seemed stunned, but then her eyes locked on me with a look of determination. She trusted me to get her out of this.

I intended to, but with the least amount of injury to her as possible. Wyatt was right. I wasn't thrilled with the idea of taking a headshot with his head so close to hers.

"What's so goddamned funny?" I demanded, my gaze dropping for a split second to Carly's waist. Her zipper was unfastened, and she was holding the waistband with one hand to hold her pants up.

Fury shot through me.

"Did you touch her?"

Wyatt's laughter continued.

"Did you fucking touch her?" I repeated, now ready to put a bullet in his head.

"No," Carly said in a panic. "He didn't! I was peeing and…"

She found a chance to escape? It made sense, but I wasn't sure what to make of her bare feet. She could have just told me that to placate me. Still, I couldn't get hung up on that possibility right now. Not when her life hung in the balance.

"Just send Carly over," I said, holding out my free hand. It was doubtful he'd oblige, but I had to try.

"Like I'm going to do that," Wyatt said, still chuckling. "I'm pretty sure I have the upper hand here. Unless you shoot me first."

"I have no intention of shooting you," I said. "Not unless you make me. I'm not into unnecessary violence."

"So you're here as Deputy Roland," Wyatt said, still catching his breath. "What charges do you plan to arrest me for?"

"How about we start with kidnapping and work our way from there?" I said, taking a step closer.

"Don't forget the murder charges," Wyatt said with a maniacal gleam in his eyes. He was a man with nothing to lose at this point. "I know you saw Carly's text." He turned to look at the side of her head. "Kudos to you for making me underestimate you."

She grimaced.

"So why'd you kill Ruth?" I asked, moving closer.

Wyatt took a step backward. "If you're expecting a confession out of me, you'll be sadly disappointed."

"It wouldn't be admissible anyway," I said, keeping the gun trained on his head just as diligently as he was keeping his on Carly's. "I haven't read you your Miranda rights."

"Then no sense wasting either of our time. Carly and I have some business to attend to, so you can go on your way," Wyatt said, taking several more steps backward and dragging Carly with him.

My index finger twitched where it curved around the trigger of my gun. There was no way I was letting him leave with her. "Carly *is* my business," I said, my voice tight with emotion. "And you're deluded if you think I'll stand back and let you take her to her father."

"Looks like you'll have to stop me," Wyatt said with a dry laugh. "And even though you have a gun pointed right at me, you still haven't shot me," he challenged. "So go ahead and *do it.*"

What was this? Suicide by cop? Then again, he knew me well enough to know I wouldn't kill him unless it was absolutely my last option.

"Just let her go, Wyatt," I pleaded, continuing to follow him and Carly down the middle of the road.

"And what?" he asked. "You'll let *me* go? I doubt that. Not with all your *evidence*."

"I don't give a shit about any of that right now," I said, not entirely a lie. "I just want you to hand Carly over so I can keep her safe."

"You're willing to ignore the files?" he spat in disgust. "You never once cared about my father's affairs. Not until Carly showed up."

"You're right," I said. "Like everyone else in town, I turned a blind eye to what he was doing. I never suspected *you* were up to anything. But taking Carly to her father won't change a damn thing that happened in Drum. You *have* to see that."

"Carly fucked up my plans in Drum, so this is plan B. I have a nice reward waiting for me in Dallas," he said. "And I'm not going to let you fuck that up too."

"I told you that my father will never pay you," Carly said, sounding pissed. "He'll lure you in and likely kill us both in one fell swoop."

"Guess we'll have to see if that's true," Wyatt said, continuing his backward progression. Up until this point, there hadn't been anyone else on the road, but tiny pinpricks from headlights shone from the top of a hill that looked to be about a half mile away. His back was to it, so he couldn't see.

My car was still running, the red lights at the back of my SUV letting them know someone was here, but there was a chance they wouldn't see Wyatt and Carly before they were too close to stop. Still, I wasn't ready to alert Wyatt to their presence yet. I could possibly use it to my advantage.

I only hoped I wasn't risking Carly's life in the process.

"Wyatt," I growled. "You know I'm not going to let you leave with her."

A slow grin spread across his face. "I don't see you trying to stop me."

He had a point. Maybe it was time to let him know I was serious.

He'd been heading toward a black van parked in an empty gravel lot. I could only presume it was his vehicle, so I swung my gun toward it and aimed at the front passenger tire and fired. Relief rushed through me when the tire released a loud pop.

Wyatt gave a sardonic look. "Glad you finally decided to man up and stop being such a pussy."

I aimed the gun back at his head. "If you equate murder and kidnapping to manning up, then I'll proudly wear the title of pussy. But now you know I'm serious about not letting you take Carly, and you also know I'm a good shot."

Wyatt hadn't moved from the center of the road, but the car behind him was silently approaching. I was going to have to decide soon.

There was no way Wyatt was driving out of here on that flat tire, which meant he was probably about to do something desperate.

He must have seen me looking past him, because he cast a quick backward glance and then turned to face me. He gave Carly a vicious shove, sending her face-first onto the ground, in the lane the car was approaching, then sprinted for the van.

Carly lay sprawled on the asphalt, and the car was only a few hundred yards away, headed right toward her.

My heart slammed against my chest, and I kicked into fight or flight. I had a choice to run after him or get to Carly, but ultimately it was no choice at all. I couldn't risk her getting run over. I ran the short distance toward her and helped her to her feet.

"Run for my car," I said. "Get in and shut the door, then call 911. Tell them there's an officer on the scene."

Her eyes flew wide. "But—

"*Do it*," I barked, giving her a push in that direction. "*Run*."

Thankfully, I heard her feet hitting the pavement as she took off, while I advanced toward Wyatt's van, my gun arm extended and pointing in his direction. I wasn't wearing a bulletproof vest, which made me a walking target.

The car started to slow down as it reached us. A loud gun blast boomed through the air as Wyatt took a shot at me. I dashed for the front of the van for cover, straining my injured thigh, which had never completely recovered from the gunshot wound I'd sustained last December.

The driver of the car hit the brakes and swerved to the other side of the road. The vehicle came to a halt, its passenger side facing the van.

Goddammit.

I waved to the driver, shouting, "Get on the highway and head to the nearest exit with an open establishment. Call 911 on the way." When the car didn't move, I shouted, "Go!"

The driver, an older woman who looked rightly terrified, got the message, peeling out and sending gravel in the shoulder flying as she headed for the I-40 entrance.

Carly was sitting in the front seat of my Explorer, the door open as she watched us.

"Give up, Wyatt," I called out. I turned back to the van, my gun upright and ready to fire. Pressing my back against the side, I inched my way down the side, glancing in the driver's window. Empty. I continued down the side until I'd reached the back. Both doors were closed.

"Give it up, Wyatt. The state police are on their way. You'll never get out of here!"

He didn't answer. Was he in the van or had he run into the brush bordering the lot?

Taking a deep breath to steady myself, I inched around the back of the van, then grabbed the handle and jerked the door open, pointing my weapon inside.

Empty, and the interior lights were off.

I felt a sudden sharp pain slam into my bad leg and saw a long metal object drop onto the ground with a tinny sound.

Waves of pain radiated through my body, and stars danced in front of my eyes. I dropped to my knees and a new jolt of pain radiated through my leg from the impact with the ground. I dropped flat onto my stomach, gravel digging into my stomach and chest.

It took me two seconds to come to my senses enough to realize that Wyatt had just attacked me and was no longer next to me. I looked up and saw him sprinting toward my Explorer, which was still parked in the middle of the road on the bridge.

Carly was standing next to the SUV, holding my rifle, and aiming it at Wyatt. Had she loaded it?

"What did you do to him, Wyatt?" she demanded.

"Go, Carly!" I shouted. "*Leave!*" She still had a chance to get in the

Explorer and drive off, but she had to run for the door now before he got any closer.

Wyatt stopped running and released a laugh. "You know she's not gonna do that. She's too damn loyal to leave you."

He was right. She didn't budge.

"Don't do this, Wyatt," I called out to him. "I know you care about her. Don't sentence her to death." I tried to get up onto my knees, but the pain made me see stars again and I fell back onto my stomach.

"I have a plan, Carly," Wyatt said, pleading leaking into his voice. "A plan to bring your father what's coming to him. But you're gonna have to get in the car and go with me willingly. The upside is that Marco stays alive."

She shot a glance at me.

Goddammit. She was going to go with him.

"Carly," I begged, trying to get to my feet again. I needed my gun. Where the hell had I dropped it?

Wyatt pointed his gun toward me. "I will shoot him, Carly. Now get in the back seat."

She gave me another long look before reaching for the back door.

I groped around on the ground for the gun. Where the hell had it gone?

"Carly!"

Wyatt stood next to the driver's door and kept his gun pointed at me. "Get in, Carly, or I'll shoot Marco. I swear."

Carly gave me one last look, then got in the back seat.

My hand connected with the handle of my handgun. I aimed it toward Wyatt and fired several times, but I hadn't taken much time to aim and my bullets didn't find their mark.

Sonofabitch.

Wyatt hopped into the driver's seat and shut the door before she shut hers. The Explorer was still running, but instead of backing up to get back on the interstate, he took off, heading south.

"Fuck!" I shouted, aiming for the tires and firing in rapid succession. He must have floored the gas pedal because it shot ahead fast enough that I obviously missed.

Sirens sounded in the distance.

The state police were now involved, and I'd be tied up here for hours while Wyatt got away with Carly. Was it time to reveal her real identity to the authorities?

A huge can of worms was about to get opened.

I reached into my pocket for my phone and realized I'd left it in the Explorer.

Dammit!

CHAPTER TEN

HANK

Max and I didn't waste any time heading to the Drummond land. He parked in front of the empty house, and we raided a closet for flashlights.

I could feel my blood sugar falling, so I grabbed some crackers out of the pantry and ate several, downing a glass of orange juice in between bites.

"You okay?" Max asked with a worried look.

"Been too long since I ate," I said. "I'll be fine."

Max didn't look convinced.

"We don't have time to lollygag," I said sharply. "We need to find the place where you think the boy is buried before Bingham and his crew show up." Then I added, "I don't feel like havin' a gun battle tonight."

"I don't feel like havin' a gun battle any night," Max muttered.

"How sure are you that the boy's buried out there?" I asked, setting my empty glass in the sink.

Max made a face. "About eighty percent, maybe ninety."

I cursed. "I was hopin' for one hundred percent, but that'll have to do."

"I could be wrong," he admitted, "but the timing fits too well.

And I remember my mom being really upset about Rodney being missing. More so than she should have been."

"Guilt?" I asked.

He shook his head. "I don't think she could have been part of killin' a kid. I think she found out after the fact."

I wasn't so sure I agreed with him, but I saw no point in arguing.

I picked up one of the large flashlights standing on end on the kitchen counter. "Let's get goin'."

We headed out the back door, walking soundlessly across the large, manicured lawn toward the trees. Max took the lead down a footworn path for a good fifty feet before it opened to a clearing.

I shined the light around us, taking in the twenty-foot by twenty-foot area. "And you have no idea where he might be buried?" I asked in a tight voice.

"No," he said, looking around. "I know she put flowers at the base of a tree, and I think it was over in this direction." He gestured to a group of trees.

"Do you remember the ground being disturbed?" I asked.

He ran a hand over his head. "I was a stupid kid, Hank."

"So that's a no," I said in disgust. "So how sure are you now?"

He looked like he was about to vomit again. "Maybe fifty percent."

Shit. I'd pinned a lot on his belief that Rodney Bingham was buried out here, which was proof I wasn't acting rationally.

I turned to look at him. "You *do* know that Bingham's gonna show up here and expect to find his brother. What happens if he doesn't?"

"He's out here, Hank. I never put it together before, but it makes sense." He leaned his shoulder against a tree. "I'm not sure why I never put it together before. She never did anything like that before or since, and I'm sure it all started *after* Rodney disappeared."

I heard a rumble in the distance, along with the sound of engines. "Well, he's on his way, so we're about to find out."

Not only did Bingham show up with a dozen men, but he'd also hauled in a backhoe on a flatbed trailer. The flatbed driver parked on the drive-in front of the house, then rolled the backhoe off and drove it across the Drummonds' lawn toward the trees, leaving deep ruts in the grass.

"Mom's gonna be pissed about that," Max said dryly, not sounding like he cared. Then again, he probably didn't. It looked like the bitch had abandoned him to deal with the legal fallout of his father's messes as well as his own. Those spreadsheets were still floating around somewhere, taken from Ruth's house, and Max's future outside of a correctional institution was floating around like a kite.

"Did you wipe those files from your computer?" I asked under my breath while we watched Bingham's men head toward us like an invading army.

The kid had been good to Carly, ever since she first showed up in Drum, and for that reason alone, I'd help him any way I could.

Max shot me a surprised look. "No. I took out the memory, destroyed it, burned it, then buried it in the woods."

Pinching my lips, I gave him an appreciative nod. Maybe the boy wasn't as stupid as I'd thought.

Max led them to the clearing, but the backhoe wouldn't fit through the trees, which didn't matter since Bingham had come prepared. Several of his men grabbed chainsaws out of the back of one of the several trucks they'd driven, and they got to work on cutting down the trees in the backhoe's path.

They'd just started in when my sat phone rang. The number was Carly's.

My heart leaped into my throat. "Carly?"

Max gave me a hopeful look.

"It's me," Marco said in a flat tone. "Good news and bad news," he said without preamble, sounding out of breath. "I found Carly, but Wyatt took my Explorer and got away with her. He now has my phone and yours. I found hers on the side of the road."

That fucking motherfucker. When I caught up to him, I was going to kill him with my bare hands. "Where are you?"

"About an hour west of Knoxville off I-40. The state police are on the way. I'll be tied up here for hours. You're gonna have to take over from here, Hank."

"I'm on my way." I hung up and went over to Max. He was standing about ten feet from Bingham, who was watching the whole spectacle with a grim face and his arms crossed over his chest.

"Max," I called out to him when I thought he could hear me. "A word."

He gave me a surprised look, Bingham also turning his attention on me.

Max walked over, looking back as the backhoe started to rumble through the new opening.

"Marco found Carly," I said.

Joy covered Max's face, but then he turned sober. "Why don't you look happy about that?" he asked in a tight voice. "Is she..."

"No," I said more forcefully than I'd intended. "He caught up to them, but Wyatt stole Marco's vehicle and took Carly with him. Now the state police are on their way, and he has to stick around and talk to them."

"What was Wyatt driving?"

"I don't know. He didn't say, and it doesn't matter. He's got Marco's SUV now. At least we know the make and model and the license number."

He gave me a long look. "If the state police are involved, is Marco gonna tell them the truth about Carly's real name and her father?"

"I don't know," I said, feeling like a fifty-pound rock had been dropped into my gut. I would have kept it to myself, but Marco was trained to trust law enforcement. If he thought it would help her, he'd do it. I just wasn't convinced he was right. "I'd bet money Wyatt's headed to Dallas, and Blakely already knows she was hiding out in Tennessee, so I'm not sure it matters." I put a hand on his arm. "I'm going to stop him."

His mouth twisted and he cast a glance back at the scene in the woods. "I should go with you."

"No, you should stay here with your eyes and ears on Bingham. Keep me informed."

He looked torn but nodded. "Okay."

"If Bingham's kid brother isn't out here, I doubt he'll hold it against you, but if you get worried, let me know. I've got my own crew who can show up as backup."

A sly grin spread across his face. "I bet you do." Then his smile fell. "Go find Carly before Wyatt does something he'll live to regret."

"Fair warning, boy," I said with a growl. "If he does something as stupid as turning her in, he won't be living long enough to have much regret."

Max gave me a long stare. "Then let's hope he's not that stupid."

CHAPTER ELEVEN

CARLY

Wyatt shot off down the road. "To say I'm disappointed is an understatement," he said matter-of-factly. He grabbed a phone connected to the car's navigation system, jerked it free from the cord, then lowered the driver's window and tossed it out onto the road.

There went my opportunity to call someone, not that I knew I'd had an opportunity until I saw it fly away. Then, to my surprise, Wyatt picked up another phone and tossed that one too.

"How many phones does he need?" he snarled, then lifted his hand between the two front seats. "Give me the gun."

"Why don't you come get it?" I shot back in a snotty tone from the seat behind him.

"Trust me. I will. Now hand it over."

I lifted the gun and pointed it at the back of his head. "Maybe I'll just shoot you instead."

"You can do that," he said, sounding amazingly calm. "But we're traveling at a high rate of speed and you don't have a seat belt on. You'll probably die in the crash."

"I'd rather die in a car accident of my own design than one of my father's."

"Don't be stupid," he said in disgust. "Every minute you're alive is another minute you can try to work things around in your favor."

"Convenient philosophy for the man who wants to turn me over *alive*."

"Necessary philosophy for a man who's had to forge his own path to get what he wants."

"And how's that working out for you?" I asked in a snide tone.

His gaze met mine in the mirror. "You've royally fucked things up, but I think I can turn them around."

"In other words, not so great."

To my surprise, he laughed. "Hand over the gun, Carly. We both know you're not gonna shoot me."

I didn't hand it up right away. Part of me wanted to pull the trigger just to prove him wrong, but he had a point. There were too many moves left on the chessboard for me to risk it all just yet. Besides, even if I were ready to die in a fiery crash, I didn't think I could shoot him in cold blood.

I handed him the gun.

He grabbed the barrel and dropped the gun down to the floorboard in front of the passenger seat, muzzle down.

I darted a glance out the back window, but we were much too far away to see Marco. My heart ached as everything that had happened over the last few minutes hit me in the solar plexus. My adrenaline crashed as my despair rose.

How had Marco found me?

The only explanation I could come up with was Hank must have put a tracker on my phone and Marco had followed it. Which meant he wouldn't be able to find me now…unless Wyatt still had my phone.

"Where's my phone?" I demanded.

He barked a short laugh. "Are you fucking kidding me? It's gone. And if I had it, I would have tossed it out seconds after we took off. I don't plan on letting Dudley Do-Right find us again."

So Wyatt suspected a tracking device too.

I stifled the urge to cry. In my head, I could still see Marco lying behind the van, desperately trying to stop Wyatt from taking me.

Was he okay? I hadn't seen Wyatt hiding around the corner of the van with a crowbar until it was too late. I'd found Marco's rifle and was loading it with clips. I'd looked up and saw Wyatt swinging the crowbar like a bat into Marco's thigh, right over his gunshot wound.

What was he thinking right now? Was he sick with worry? I knew he was probably blaming himself for Wyatt taking me again, but he wasn't to blame. I only wished I could tell him so.

I had to find a way to get Wyatt to see reason.

"The state police are on their way, you know," I said. "They're going to come looking for us. You might as well turn yourself in."

"Good try," he said with a dry laugh. "I think I'll pass."

"What if they find you?"

"You mean *us*. You better hope they don't because I'm not planning on letting them haul me in, which means there'll be a shoot-out, and the likelihood of you getting shot is pretty high."

"You're an asshole."

"I know."

We drove in silence for a few minutes, then he turned down a side road.

"Where are you going?"

"You don't need to worry about that," he said. "Just sit back and enjoy the ride."

"Don't you try to patronize me! I'm sick of everyone expecting me to go along for the ride. I'll likely be much more cooperative if you just tell me what you're up to." Now that I'd tried to run away, playing the dutiful captive was a wasted effort. I might as well be the real me.

"That's bullshit and we both know it."

"Okay, I'll be a *little* bit more cooperative if you tell me."

He grinned. "As you so logically pointed out, they're gonna be headed this way, looking for us. So I put some distance between us and them, then headed down a side road. I plan to take a few turns and find a place to get a few hours of sleep, then we'll take the back roads for a bit before making our way to Texas."

"Aren't you worried about making your deadline?"

"It's a loose schedule."

Which meant he hadn't made an appointment with my father yet. That filled me with more relief than it probably should have. It only meant I had a little more time left before I met my demise.

True to his word, Wyatt made several more turns, following side roads using the GPS.

"When'd you get a smartphone?" he finally asked after about a half-hour of silence.

I pressed my lips together and remained silent.

"Marco get it for you?" he asked with a sneer. "You looking for a man to take care of you?"

"I can take care of myself, and you damn well know it. But *no*, Marco didn't get it for me. It's worthless in Drum."

"Exactly, it's worthless in Drum, so why have it? Were you planning on leaving?"

"Why the fuck do you care, Wyatt?" I asked with a sigh. "What difference does it make?"

"Where were you planning on going?"

The silent treatment was my best option, but I couldn't help needling him. "Maybe I didn't need it for where I was going so much as where I'd *been*."

"That's right," he said with a sneer, glancing at me in the rearview mirror. "You've been *traveling*. You made a trip to Chattanooga and got a copy of the tape of my father confessing to murder."

"Murdering your baby sister," I said. "*Ruth's* baby sister."

He didn't respond.

"That made you and Ruth family."

He remained silent.

"Did she know you were paying Franklin to spy on me?"

He didn't say anything, and I was sure he wasn't going to, but several seconds later he said, "No."

"Did she find out? Is that why you killed her?"

"I didn't kill her," he said quietly.

"I don't believe you. Just a few hours ago, you told me you did. You've admitted it since."

"I lied. I was trying to scare you and convince you to believe I'd hurt you."

"Well, you succeeded. If you didn't kill Ruth, who did?"

"Franklin."

I shook my head. "No. *No way*. He loved her."

"He didn't mean to," he said, his voice breaking. "It was an accident."

"I don't believe you."

He shrugged. "I can't make you believe me one way or the other. Besides, I didn't pull the trigger, but I suppose I was the reason for the argument that led to her death. So, in a way, I *did* kill her."

I let his words sink in. "You saw Franklin kill her?"

"Ruth found out that Franklin was spying on you and reporting to me. She was furious. She also realized that no one had run him off the road. He'd made it up. It was his fault, so his insurance wasn't paying for it, and she was furious about that too. I'd never seen her that mad before. He found her spreadsheets and told her if she didn't calm down, he'd take them to Bart, which pissed her off more."

Tears sprung to my eyes. "I still don't believe he killed her."

"Like I said, it was an accident."

"How'd you happen to be there and see it all?" I asked suspiciously.

He grew subdued. "I dropped by to tell Ruth about her mother's baby."

"Your sister," I sneered.

His voice broke. "*Our* sister."

My brow lifted in surprise. I hadn't expected him to be

emotional about it, but then again, he'd been protective of Lula when he'd found out she was his half-sibling.

He rubbed a hand over his face, then dropped it. He looked exhausted as he locked gazes with me in the mirror. "I was collecting evidence, Carly. Evidence to bring the fucker down. I asked you to trust me."

"If you were collecting evidence, then why were you so pissed that your father was arrested?"

"Because this is all so much bigger than a baby's murder, Carly!" he shouted.

I stared at him in disbelief. "What on earth could be bigger than the cold-blooded murder of an innocent baby, Wyatt?"

He didn't answer.

He drove for several more minutes before turning onto a gravel road tucked between a grove of trees.

I sat up on the edge of my seat. "Is this safe?"

He released a bitter laugh as he drove down the overgrown one-lane drive. Trees surrounded us on both sides, and it looked like the road ended about forty feet ahead. "Safer than running from the police."

"News flash, we're still running from the police."

"Technically, we're hiding from the police, and I seriously doubt they'll find us here."

He slowed to a halt and put the Explorer in park, then shut off the lights. We were plunged into such deep darkness I could barely make out his features.

"Now what?" I asked.

"Now we get some sleep and head to Texas in the morning."

He scooted his seat back all the way, then lowered it down until he was practically lying on my lap.

"Hey!" I shouted, moving over to the other seat.

"You're welcome to sleep wherever you like." Then he added, "As long as it's in the car."

He didn't plan on tying me up? Not that I'd seen anything he could use to do so. "It's going to get hot and stuffy in here."

"I'll crack the windows open a bit, but if we lower them too much, we'll get eaten by mosquitoes."

"I have to pee," I said with plenty of attitude.

"Then get out and pee."

My mouth gaped. "You're going to let me get out by myself to pee?"

My eyes were adjusting to the darkness, and I could see him shrug. "Why not? We're on a gravel road and you're barefoot. Even if you take off running, I'd catch up to you in seconds. So go for it. I could use another run."

I stared at him in disbelief, sure this was a trick, but he crossed his arms over his chest, and I could barely make out his eyelids closing.

He was going to let me get out of the car on my own. "This is some kind of trick."

"It's not a trick," he said, his eyes still closed. "Pee if you like, then get back in the car. Unless you want to get eaten by mosquitoes."

I gave him one last stare before I opened the back door. He must have turned off the interior lights because they didn't come on. I carefully stepped out of the car, positioning the ball of my foot so it wouldn't get stabbed by a piece of gravel.

"Got any tissues?" I asked.

"You'd know better than me," he said with a sneer. "It's not my car."

I made a face, even if he couldn't see it. "Center console."

Grumbling, he opened it and pulled out a few tissues, then closed the lid and leaned back to hand them to me. "You must have domesticated Marco. He doesn't seem the type of guy to have a travel-sized tissue box in his truck."

"Shows what you know," I said as I snatched them, leaving the door open as I made my way to the back of the Explorer. He'd had them in his car when we'd teamed up looking for Lula last December.

"Shut the door," he shouted.

"Do it yourself," I blurted back, taking a tentative step on the side of the one-lane road. Dried grass and sharp sticks stabbed the bottom of my foot.

Escape through the woods was out.

Pulling my pants down, I squatted next to the back bumper and peed on the gravel. No way was I attempting to pee in the brush. My gaze took in the end of the gravel road, which I could barely see in the darkness. Could I make a sprint for it? Should I risk running through the trees anyway? But everything was dry and crackly, so Wyatt would find me if I risked the woods, and if I tried the gravel road...he'd catch me before I made it to the highway.

I finished peeing and wiped before I stood and fastened my pants. I knew Marco had a plastic bag for my trash in the pocket behind his front seat, but I decided to leave it on the side of the road, under a small rock that might hold the tissues in place if a wind blew through. DNA evidence. There was a slim chance anyone would think to look here, but I planned to take advantage of every opportunity I got...

"Shut the damn door already," Wyatt shouted, sounding more cranky than pissed.

I really didn't want to get back inside with him, but I could already feel several itchy welts on my arms and one on my butt, so I got back in the car and shut the door.

"Happy now?" I snapped.

"Peachy."

I tried to lean the back bucket seat on the passenger side back, but it didn't budge. If I had to sleep in here, I was getting comfy, so I climbed into the third-row seat and leaned over the back, finding the buttons that lowered one side of the back seats.

The suitcase I'd abandoned when I was kidnapped was in the back, so I unzipped it and grabbed a sweatshirt to use as a pillow.

Balling it up, I laid down the passenger side of the second and third-row seats, then stretched out with my head toward the front.

I lay like that for a few moments, my mind drifting to Marco lying on the gravel behind the van, in agonizing pain. Wyatt could have shot him. I supposed I should be grateful.

Grateful he'd hurt Marco? How messed up was that?

"How did Franklin kill Ruth?" I asked softly.

Wyatt didn't answer right away. In fact, I thought he might be asleep, but he shifted in his seat and pushed out a long breath. He scrubbed his face with his hand. "Ruth had a gun. They fought over it."

Tears stung my eyes. "And you just watched?" I didn't try to keep the bitterness out of my words.

"It all happened so fast," he said, his voice breaking. "I still can't..."

I believed he hadn't killed her, but it still didn't excuse all the other crap he'd pulled, not the least of which was kidnapping me at gunpoint.

He cleared his throat, shifting in his seat again. "I've had that video of my father talking to Patricia, the owner of that bar in Chattanooga, for a while. The one that shows my father admitting he'd murdered his baby. But I'd only recently figured out who it was in the video."

"How'd you put it together that the baby was Ruth's sister?"

"It didn't happen all at once. The wheels started turning after Max and I discovered Lula was our sister, and Max began to wonder if we had more siblings out there. It became obvious that he was the one who'd tried to kill her, not the man who'd raised her, and then..." He rubbed his forehead. "I not only started to wonder how many siblings he'd sired, but how many he'd *killed*."

"Did you ever come up with a number?"

He was silent again, and when he spoke, he ignored my question. "I traced the murder to Ruth's mother by accident. One night I was half drunk, thinking about when I was in my early twenties and acting like a fool with my friends—"

"I guess not much has changed besides the friends part," I blurted out before I could stop myself.

He turned his head to catch my eye. "You're telling me I'm a fool?"

I lifted a brow, even though I wasn't sure he could see my face well enough to catch it. "If you're willing to own it..."

"You have no idea what I've been doing, Carly," he said with plenty of bitterness.

"And whose damn fault is that?"

He crossed his arms over his chest again, fuming for several seconds. "You think you're so goddamned perfect."

"I've never *once* claimed to be perfect."

"Why couldn't you just trust me?"

I pushed up on my elbow and stared into his shadowed face. "You kidnapped me at gunpoint, Wyatt! You held a gun up to my head and threatened to shoot both me and Marco. You just beat Marco in his injured leg with a crowbar. Why the hell would I ever *consider* trusting you?"

He studied me for several seconds, then stared up at the ceiling.

"Just let me go, Wyatt. You can drop me off at a gas station somewhere, and I'll call Marco and leave the police out of it."

"Like I'd believe that," he scoffed. "The police are already involved."

"You of all people know how much I've tried to stay under the radar. My father knows that I was alive and well as of yesterday morning. You can bet your ass he'll be looking for me, and if my name turns up in a police report..."

"Don't you think Marco's gonna give your name to the police? It's already a done deal," he said bitterly. "You're better off sticking with me."

"Ha!" I barked. "Do you think I'll have an easier demise if I let *you* turn me in?"

He turned onto his side to face me. "I was never going to turn you over to him. I only said that to punish you for screwing everything up."

I slowly shook my head. "That's utter bullshit."

He shrugged. "Believe what you want, but it's true. You're bait."

"Bait for what?"

He leaned back in his seat again, closing his eyes. "Get some sleep. Tomorrow's going to be a long day."

Little did he know. I planned to make it pure hell.

CHAPTER TWELVE

MARCO

The state police spent hours working over the crime scene. Of course, they had questions about why Wyatt would want to kidnap Carly. I nearly told them the truth, but something stopped me from sharing her real identity.

Maybe it was because they'd be skeptical. Randall Blakely was a highly respected man, and while the FBI had been interested in him, his involvement in the Hardshaw Group hadn't been widely shared with the general public. If I started spouting off about Hardshaw and Randall's missing daughter, they might think I was a conspiracy theorist. I needed them to take me seriously. I needed them to look for her.

I'd seen the inside of Wyatt's van. There was plenty of evidence he'd tied her up and duct-taped her mouth.

What was he doing to her now?

I started to spiral into hopelessness, but I reminded myself that we had more information to find him now than we'd had before. The fact he was driving my Explorer gave us an advantage, but I suspected he'd dump it sooner rather than later.

One more reason not to tell the state police about her real identity was that it could potentially get her in trouble. While she'd never tried to use her identity for anything other than opening a bank account, she could still be charged for using a false identity, a misdemeanor in Tennessee. And if I told them that I'd known for months, I suspected my career in law enforcement would be over.

Then again, given the issues I'd been having with my leg, I already knew it was over.

They'd wanted to call an ambulance for me to get my leg checked out. I had to admit Wyatt had done a number on it, setting me back weeks—possibly months—of progress, but there wasn't anything an ambulance or an ER could do for me. All they'd offer was hardcore drugs, and I'd refuse those. I needed my wits about me.

But the biggest reason I hadn't told them was her father. If he caught wind that she'd been kidnapped, I was sure he'd find a way to use it to his advantage, and I wasn't there to protect her.

Still, I couldn't help questioning my decision. Had I just helped or hurt her?

Hank arrived as the police were wrapping up their investigation. I was standing on the side of the road, across from Wyatt's van, when he pulled his car to a stop on the side of the road next to me and got out.

A policeman walked over to tell him to move on, but I held up a hand and said, "He's with me."

Hank swung his crutches quickly to reach me, then turned his gaze on the van.

"How'd you find me?" I asked. "Do you have something else to track her phone?"

"Nope. Saw all the commotion from the highway."

I gave a sharp nod. "Well, he has both of our phones, so we're screwed."

"We'll find another way." He sounded so sure of it that I almost believed him.

I was glad Hank was here, but somehow his presence loosened the vise that had been holding me together.

"So that's what he used, huh?" Hank asked, flicking a hand toward the van.

"Yeah," I said, the exhaustion creeping into my voice.

"Do you have to stay here?"

"They haven't told me I can go," I said.

"Are you in trouble?" he asked under his breath.

I released a sharp laugh. "Not in legal trouble."

"Then you're fine," he said gruffly. "Everything else can be dealt with."

A lump filled my throat. "She was right here, Hank. She was on the road when I pulled up, and I was *so fucking close* to getting her." My voice broke. "But he…" I stopped, unable to continue.

"Did he hurt her?" Hank asked in a dark tone.

"Not in front of me. He held a gun on her…" I took a deep breath and tried to contain my emotions while I told him what had happened.

He studied the van while I talked, his jaw tightening when I told him about the evidence in the back of the van.

"So he was *planning* to take her," he said. "It wasn't a spur of the moment decision."

"Looks like it."

Balancing on his crutch, he lifted his right hand and rubbed his stubbled jaw. "You got any kind of tracking device in that car of yours? Like that OnStar shit?"

"No, but the police can get a warrant to track the GPS. I could find them with the app on my phone, but like I said, Wyatt took both our phones, and I can't get a new one until a phone store opens."

"And like *I* said, we'll find another way. We need to find them before the police do."

"How do you plan to do that?"

"Head them off at the pass. We're going to Texas. Did you tell them why Wyatt likely took her and what he plans to do?"

I ran a hand through my hair, close to breaking down. "No. But now I'm second guessing that decision."

"You did the right thing, son," he said, nodding. "I doubt they would have believed her father wants her dead. They would've written you off as a nut job. Besides, they'd have taken her straight to her father if they found her. This way, unless Wyatt opens his big mouth, they'll return her to us."

I nodded, still not sure I'd handled it right, but said, "Yeah."

"So that's done," Hank said. "Now we need to focus on how to get her back."

"We need new phones with our old numbers," I said. "She has no way to reach us if she finds a chance to call."

"Agreed, but there's no guarantee she'll try callin' you anyway. Not if she knows Wyatt took our phones."

I nearly fell to my knees. I'd blown my opportunity to save her because I'd been careless. I should have just left him with his van and taken Carly far away from him. Instead, I'd wanted to make him pay. Up until now, I'd been too adamant about staying within the lines. Not anymore.

I was saving Carly, no matter what it took.

CHAPTER THIRTEEN

CARLY

I struggled to fall asleep. For one thing, I was uncomfortable. The folded-down seats were hard. I could've climbed into the front passenger seat and tried to sleep there but doing that would feel like giving Wyatt another point in this weird game we were playing, and I was already losing by a mile. So I ignored my stiff joints and focused on coming up with a plan for once we left this hidey-hole.

Presuming Marco hadn't taken anything out of my bag, I had a pair of shoes I could wear to escape. Wyatt had disengaged the interior lights, so could I quietly open the door, sneak out, and run? I couldn't remember the last time I'd seen a house on this godforsaken road, but I could run and hide in the woods.

It was the best plan I had.

I decided to test it before I actually ran. Reaching for the door handle, I gave it a slight tug, gently pushing the door open. The interior lights instantly came on, along with a chiming bell.

"Going somewhere?" Wyatt asked dryly. He was on his side facing the driver's door, his back to me.

"I have to pee again."

"Okay," he said in a challenging tone. "Go for it."

I got out and squatted by the bumper again, but I'd hardly drunk anything, so not much came out.

When I got back inside, Wyatt mumbled, "Get that out of your system?"

I didn't answer. We both knew he wasn't talking about me peeing.

Obviously, all his kidnapping supplies had been abandoned along with his van, but I got the impression he wouldn't have tied me up anyway. It was like we'd moved to a new level of our kidnapper/kidnappee relationship. There was no denying I'd had the opportunity to shoot him on that bridge, yet I hadn't.

After all the shit he'd pulled, I still couldn't kill him in cold blood. I wasn't sure I could shoot him at all.

I wasn't so sure that wasn't a flaw.

I had to admit I was still curious about a lot of things, from what had really gone down with Ruth to what he'd intended to do with that secret basement room. And, of course, what he planned to do with me if he wasn't going to hand me over to my father.

I finally fell asleep, and the next thing I knew, Wyatt was calling to me.

"Wake up, Sleeping Beauty," he said. "You have three minutes to pee before we leave."

With my eyes half closed, I grabbed my suitcase and dragged it up close to me, then dug inside for my toiletry bag.

"What are you doing?" he grunted, twisting in the seat.

"Finding my toothbrush." I triumphantly pulled it out. "Marco had my bag in the back, and I'm not going anywhere until I brush my teeth." I glanced around. "What time is it?"

"Nearly six."

I climbed into the seat behind the driver's seat and noticed a bag sticking out from under it. It had a couple of bottles of water inside. I took one out and opened it, then took a few long gulps.

When I lowered the bottle, I noticed Wyatt watching me with what looked like longing, only I suspected it wasn't for me.

I handed him the bottle. "Save some for me to brush my teeth." I didn't expect him to comply, so I was surprised when he did.

I got out and peed again, a wasted effort since there really was nothing left, then put toothpaste on the toothbrush and began to brush.

Wyatt opened the driver's door and slowly got out. "We need to talk about expectations for the day."

"Good idea. I expect breakfast soon," I said around my toothbrush. "I'm starving."

"It's not my fault you didn't eat much last night," he countered with a dark look.

Holding my hair to one side, I leaned over and spat toothpaste on the ground. I shot him a glare as I straightened. "You kidnapped me at gunpoint, tied me up and gagged me, then told me you were turning me over to my father. Sorry if I didn't have a hearty appetite."

His mouth twisted to the side. "Time to go. Get in the car."

I stuck my toothbrush back in my mouth and finished scrubbing.

Wyatt looked torn between throwing me in the car or flat-out strangling me on the spot, but he settled for some more glaring. When I finished and swished what was left of the water in my mouth and spat, he snatched the toothbrush and paste out of my hand and then started to brush his own teeth.

I made a face. "Gross."

"It's not like our mouths haven't touched before," he said with a leer.

I made another face, adding plenty of disgust. "I don't need the reminder." I especially didn't like *him* thinking about us kissing.

I briefly considered making a run for it, but I was still barefoot—an idiot move on my part. Still, even if I'd thought to put on shoes, I didn't have a realistic shot at getting away. Wyatt would have caught me in seconds. I considered trying to lock him out of the car, but I could see the key fob dangling out of his pocket, taunting me as he paced the side of the gravel lane.

So I resigned myself to staying with him and opened the back doors.

Wyatt shook his head, then pointed to the front of the SUV as he continued brushing.

"You want me to sit in front?"

"Best way to keep an eye on you," he said around the brush.

I narrowed my eyes, trying to figure out if he was serious, but I wasn't ready to comply with his demands. My shirt stank of BO, and while it might be great revenge to make him smell me, I didn't want to punish myself too. "I'm going to change my shirt first. Stay up there while I go in back to do it."

I didn't wait for permission, just walked around the back, carefully picking my way over the gravel pieces. I opened the hatch and dared to glance up at him through the front windows. He looked aggravated but continued brushing. Could I find something back here to use against him? Maybe there was something I could use to hit him over the head or surprise him.

I rummaged through my bag, but the heaviest things were a book and a pair of shoes. Neither would do much damage to a man as big as Wyatt.

"What are you doing back there?" he called out.

"I'm trying to figure out what to wear."

"This isn't a damn fashion show. What's taking so fucking long?" he grunted as he rounded the corner and found me digging in my bag.

"If I'm meeting my dad tonight, I don't want to look like I'm homeless."

He pushed out a sigh, and some of his attitude faded. Resting his butt against the edge, he said, "We're not meeting him tonight, Carly."

So I had at least twenty-four hours. I had to make the most of it.

Then, to my surprise, he added, "And you could never look homeless."

I pulled a gray scoop-neck T-shirt from my bag and turned my back to him, then tugged my shirt over my head and pulled on the new one. When I turned around, I found Wyatt staring at me, which meant he hadn't looked away. Not that I cared much. Even if he'd seen my bra, it was more modest than some bikini tops.

I tossed my old shirt on the floor of the cargo area and dug deodorant out of my toiletry bag. Lifting the edge of my shirt, I applied it to each armpit, wishing I had wet wipes or something to clean up with first.

Wyatt continued to study me, then stood. "I take it the other bag is Marco's? We're about the same size." He unzipped it and started to rummage around.

Wyatt was a little more buff than Marco, but I had no doubt Marco's T-shirts would fit him. Part of me wanted to stop him as I watched him pull out Marco's favorite T-shirt, a souvenir from a concert he and Max had gone to in college. It seemed wrong for him to wear it, but I bit my lip and sat on the back so I could put on socks and a pair of athletic shoes I'd packed in the bag.

"Expectations," he said in a dark tone, the expression on his face matching.

I stared up at him, waiting with a dry expression.

"You do realize that I've kidnapped you, which means I make the rules," he growled.

"So make the rules."

He still looked exasperated.

"Are you wanting me to cower with fear, Wyatt? Because I'm not going to do that today." The Wyatt who had taken me yesterday had scared the shit out of me, but up until this point he hadn't hurt me, which gave me courage to stand up to him. Still, I'd be stupid to push my luck too far.

"Look," I said, purposely softening my tone. "I'm stuck in this situation, and I have no desire to be shot, so just tell me what you want so I know what to do."

Some of his edge faded. "I'd rather not tie you up today; instead, I'd like to appeal to your logical side."

Tilting my head, I studied him. I suspected he wanted the reward, but it didn't hurt to clarify. "At one time, we had similar goals. If you tell me what it is you want from my father, maybe I can help you achieve your goal."

"Why would you do that?'

"Um, to save my own skin?" I shot back sarcastically. "So what do you want? The reward?"

He was silent for several seconds, then turned and started to walk away.

Dammit. I'd pushed too hard too soon. But he surprised me by returning and leaning against the back of the Explorer again. "Money. I need money."

"So your plan is to turn me in and collect the reward?"

"No. My plan is to collect evidence and then turn it over to the FBI."

My jaw literally dropped before I recovered enough to say, "*What?*"

"What do you think I've been collecting all that shit on everyone in town for?" he asked in disgust. "To turn it over to the FBI."

I shook my head, trying to make sense of his words. "Wait. *What?*"

"Look," he said in a pleading tone, "I was trying to bring my dad down, just like I told you after you came to town, but I didn't want to involve you in it. It was too dangerous."

I studied him, still stunned as I tried to sort through everything, I knew about him through this new lens.

"Remember when I left town to go buy a tow truck in West Virginia last year?" he asked. "A few weeks after you'd come to town, I set up my first meeting with the FBI. I was trying to work out a deal."

"Why would you need a deal? What did you do that was illegal?" Then I added, "Other than the obvious of associating with Louise and Derek and getting Jerry killed."

"I didn't get Jerry—" he started to protest, then stopped. His gaze narrowed, and he waited a moment before saying, "My dad still has some money, not to mention the house, land, and the resort. I didn't want it all to be confiscated when they busted him. They agreed that I'd get to keep the properties if I got the goods to arrest him."

"So why not tell me from the start?" I asked, not sure I was buying this.

"Because that was part of the deal," he said. "I couldn't tell anyone. Which is why I tried to keep you out of all of it."

"Because you wanted his money," I said in disgust.

"Not just for me, Carly. I wanted it for my brother and my mom too."

"Your mom is just as dirty," I snorted. "And she used your poor brother."

"I didn't know that then," he said. "I only found out about the money laundering a few days ago."

"Didn't you launder money for her when *you* ran the tavern?"

"What? No!" he said emphatically. "She started that after she convinced my dad to give the place to Max. She knew I'd never go for it. But Max...well, let's just say he's a momma's boy, so it's no surprise."

I wasn't so sure I believed that either, but I saw no point in arguing with him over it. I had bigger things to worry about.

"You worked with Louise and Derek."

"To get evidence on my father. The FBI knew what I was doing."

"You got Jerry killed. Poor Jerry who defended you," I said, my voice breaking. "That man loved you and you..."

"I know," he said, his voice breaking. "I've made some mistakes along the way, but I tried to get Jerry to back off and let me handle it"—his voice hardened— "but he was too damn stubborn."

"You were paying Franklin to spy on me!"

"To protect you, Carly," he pleaded, turning to face me. "I was getting so damn close to getting what I needed, and Louise and Derek hated you snooping around. I needed to make sure you were safe."

"You claim to care about me, yet you've accused me of colluding with Todd Bingham at every turn."

He stood, turning to face me as he held out his hand. "I was jealous, Carly!"

"Of Todd *Bingham*?" I asked in disbelief.

"I was stupid, okay?" he said defensively. "I figured you'd grown impatient with me and turned to Bingham for help with your quest to bring down my father. That made the two of you dangerous. I worried you'd get in the way of my fact-finding mission."

I studied him with narrowed eyes. I wasn't sure I believed any of it. "You claim what you were doing was more important than a murdered baby. What were you looking for?"

"Extortion. Racketeering. Bribing government officials," he said in disgust. "Trust me, murder was just the tip of the iceberg."

"Murdering a baby's a big fucking deal, Wyatt," I said in disgust. "A bigger deal than the rest of it."

"Maybe so," he said with a hard glint in his eyes, "but somehow the government cares more about the other shit."

I didn't answer, because even if he was lying about working with the FBI, I suspected he was right about that.

He continued to watch me for a few seconds while I tried to make sense of his claims.

"The FBI contacted me after my father was charged with murder. They said the work I'd been doing for years was wasted effort," he said bitterly. "*They* wanted to bring charges against him, not compete with the state police. My deal was off."

"So then you decided to *kidnap* me?" I asked, my anger building.

He ran a hand over his head, looking lost. "I handled it wrong; I'll freely admit that."

"You *think*?" I shouted. "You threatened to kill me, Wyatt! You threatened to kill other people!"

"I never would have done it, Carly!" he pleaded, then hesitated before his tone softened. "You know I care about you. A lot." He took a step toward me.

I hopped off the back of the Explorer and backed away from him toward the trees. "Oh, hell no. You can't claim that. Not after the last couple of days. Not after what you did to Ruth."

"I didn't kill Ruth. I *told* you that."

My jaw hardened as I shook my head in disgust. "After you told me that you did."

"I already told you that I only said that to scare you so you'd come with me. I wasn't thinking straight. I was pissed."

"Pissed?" I shot back. "Try insanely furious! You tied me up, Wyatt! You held a fucking gun to my head!"

"I wouldn't have shot you, Carly! I could have shot you back there with Marco and I didn't."

"Only because you think my father will pay you more if I'm alive."

"Yeah, I need you alive for my plan," he grudgingly admitted, "but I also need you alive because I can't let anything happen to you." He closed the distance between us, then wrapped his arms around my back and pulled me to his chest. "I'll make sure you're safe, Carly," he said in a whisper. "I swear it."

Tears stung my eyes. I pushed against his chest, trying to break free of his hold, but his arms stiffened to keep me in place.

"I know you're hurt and pissed, and I know you hate me right now," he said in a soothing tone, looking down at me. "I don't blame you. You have every right. I also know you love Marco, and I respect that. Carly," he said. "I could have killed Marco back there on the road, but I didn't. Even though it would have saved me a hell of a lot of trouble. Doesn't that prove that I'm not the monster you think I am?"

He had a point. He could have killed Marco, but he must have known I wouldn't cooperate with him if he had. So had he spared Marco because he wasn't a cold-blooded killer or because it didn't fit with his plans for me?

"All I'm asking is for a chance to prove that I'm not going to hurt you."

"By letting you take me to my father?" I asked sarcastically.

"I'm not handing you over," he said. "I could never do that to you. I told you. This is Plan B, which I had in my back pocket all along. I figure if I make arrangements to hand you over for money and then get your father to admit he's planning to kill you, then we'll have the evidence the feds need to arrest him."

"That's not a federal crime," I said.

"Trust me, they're interested anyway," he said emphatically. "They confirmed it yesterday."

I pushed hard against his chest and broke free. "This is bullshit, Wyatt! You know I want to deal with my father! Why kidnap me at gunpoint? Why hogtie me in the back of your van? Why not just tell me what you were planning and ask me to help you?"

"Because I knew you'd never go for it," he said, anger tinging his words. "You'd insist on Marco being part of it, but he's in law enforcement, and I don't trust him."

"What has he ever done to make you not trust him?" I countered.

His lips pressed shut.

"Yeah, exactly," I said, walking a few steps away from the van toward the highway several hundred feet away. "Nothing."

"Okay, I'm jealous," he grunted. "Happy now?"

I spun around to face him. "So you let your jealousy override *my protection*?" I fisted my hands at my sides. "If you really care about me like you claim to, then you would have included him, if for no other reason than to have one more set of eyes to make sure nothing went awry."

"Well, it's a done deal now," he said, throwing his hand up at his side. "We'll have to make the best of the situation we've been dealt."

"You mean the situation *you* dealt for us," I countered bitterly.

His eyes darkened. "If you had just left well enough alone, none of us would be in this situation."

"And there he is," I said triumphantly, gesturing to him. "The real Wyatt Drummond. Blaming other people for his messes."

He stalked toward me. "I never claimed to be a saint. In fact, I'm pretty damn sure I've let you know right from the beginning that I'm not one."

"Yeah, well, stupid me for not believing you."

He glared down at me. "I'm an asshole, Carly. I won't deny it, but I'll do everything in my power to make sure you're not hurt while getting the information we both need to bring your father to justice."

"It's not too late to call Marco to—"

"Are you fucking kidding me?" he asked derisively. "Even if I wanted to ask him to help, there's no way in hell he'd agree to help me now."

"We can at least try!"

"No!" he barked. "We're not calling him. He's too by the books. You and I both know we can't follow the books to get what we need."

I turned away. While I didn't like it, I suspected he was right.

"So what's your plan?" I asked.

"Head to Dallas. We'll scope out his house and his office, then plan from there."

I chewed on my bottom lip, running through my options. While Wyatt might be telling me the truth about his end goals, there was no way I wasn't going to contact Marco the first chance I got. Still, I needed to make Wyatt think I was going along with his plan, and I had just the carrot to dangle in front of him.

"We need to make a stop before we get to Dallas." I turned to face him. "We need to go to Arkansas."

CHAPTER FOURTEEN

MARCO

After the state police let me go, Hank and I got into his car, and I cast a long look at the road Wyatt and Carly had headed down.

Hank cast a long glance at *me*. "You want to follow them?"

It felt wrong not to try, even though the state police had already searched in the hopes of intercepting them. Wyatt was too smart to get caught so easily. There was no telling where he'd gone. Trying to chase after them would be wasted time and effort.

I shook my head. "It would be pointless. We need to head to Texas."

"It's damn near two in the morning, and we're both beat," Hank said he pulled away from the curb and headed for the entrance ramp. "I say we get closer to Nashville, find somewhere to get a few hours' sleep, then take off to Dallas first thing."

We both needed sleep—Hank was looking rough—but I struggled to get on board with his plan. I'd already lost over two hours dealing with the state police. I couldn't afford to lose more time.

"You're no good to her if you're not on the top of your game," he said, keeping his gaze on the road as he merged onto the highway.

"Wyatt's a tricky one. We're both gonna need our wits about us to trap him."

"Yeah," I said, laying my head back against the headrest and trying to stretch out my leg. Wyatt had really done a number on it, and I was worried it would impede my ability to help Carly when the time came.

A wave of exhaustion and helplessness washed over me. I'd never felt so defeated in all my life. I'd failed to save her.

"We'll get her back," Hank said, his voice thick with emotion. "I swear on my own life, we'll get her back."

I didn't respond, instead closing my eyes and letting my encounter with Wyatt replay in my head for what had to be the hundredth time. I should have been more careful. I should have gotten into that car with Carly and taken off.

I should have—

I could fill my head with a million should-haves, but it wouldn't change the fact that I could have saved her, and I hadn't.

If anything happened to Carly, it was on my head.

We hadn't driven long before Hank's satellite phone rang in the back seat. "Grab that," Hank said. "It might be Max."

Somehow, in all this mess, I'd forgotten about Max. "He wouldn't come?"

"He's dealing with a big ole pile of shit back home," Hank grunted. "Now get that before he hangs up."

I unbuckled my seatbelt and pain shot through my leg as I leaned over the seat to grab the phone in the back. Hank reached over and pushed a button at the top. "Chalmers."

"It's Max. Any news on Carly?"

"No," I said, trying to keep the defeat out of my voice. "And Wyatt has a head start of at least a couple of hours."

"How are things goin' there?" Hank asked.

"That's why I'm callin'," Max said, his voice breaking. "They found a body."

I sat upright. "A body? *Franklin?*"

"No," Max said quietly. "Rodney Bingham."

I scrubbed my face. "Wait. *What?*"

"We found out what Bingham was searching for and why," Max said. "He was trying to prove that my dear old dad was Rodney's father...and that he killed him too."

My mouth dropped open, and I turned to Hank, who didn't seem the least bit surprised. "You knew about this?"

"Not until tonight. But Max was with Bingham and his crew, who were diggin' in the woods behind the Drummond house, when I left to come to you."

I ran a hand over my head, trying to piece this together. "How'd Bingham know to dig behind the house?"

Max told me about his mother putting flowers out there after the boy's disappearance and went on to explain how Bingham had found out the truth about Rodney's parentage.

"So what's Bingham planning to do with this?" I asked.

"I don't have a clue," Max said. "I asked him if he wanted to call the sheriff, but he just had his men scoop up the body and take it with them."

"He needs to report it," I insisted.

"Yeah, well," Max said wearily, "it's Bingham. I'm not going to make this the hill I literally die on."

"How does he know for sure it's Rodney?" Hank asked. "We know Bart's killed at least one other kid and tried to kill Lula. It could be someone else."

"He's certain," Max said. "Rodney's school backpack was buried next to him, plus he was wearing the clothes he was last seen in. They were pretty raggedy, but the stripes on his shirt..." His voice broke, and I heard him take a deep inhale. "I suggested that Bingham do a DNA test, but he told me that he'd take it from here and to forget they were ever there." He released a sharp laugh. "Forget my father killed yet another one of my half-siblings."

"How're you doin' with all of this?" I was worried about him. He was back in Drum all by himself, dealing with heavy shit.

"I'm too damn tired to even sort through it. I'm gonna stay in my old room at my parents' house and then start going through my mother's things first thing in the morning. Maybe I can figure out where she's hiding."

"Yeah," I said, "Good idea."

We were quiet for a moment before he said, "So Wyatt really kidnapped Carly, huh?"

"Yeah," I forced out. I knew it had to kill him to get confirmation. "I'm sorry, Max."

"Not as sorry as I am," he said, his voice breaking. "I had no idea he'd ever try anything like this Marco, I swear."

"Of course you didn't," I said, trying to keep it together. "We all thought he was sketchy, but we never expected anything like this." I clenched my fists because while Max needed to know this next part, it was really going to hurt him. "There's one more thing, Max." I hesitated, hated that I was the one to tell him. Hated that I wasn't there to comfort him. "In Carly's text, she said that Wyatt killed Ruth."

"No," was his mangled response.

"I'm so sorry," I choked out.

We could hear his ragged breathing for several long seconds, then he sniffed. His voice was tight when he said, "Yeah, well, we'd figured out my brother is a piece of shit, and the evidence just keeps on piling up to prove it."

"I'm sorry," I whispered.

"Not half as sorry as I am." He took a moment, then asked, his voice stronger, "Are you okay? Hank told me about your leg."

My face flushed with shame. My leg throbbed, and I could feel my thigh muscle swelling against my jeans. "I'm fine."

"He's not," Hank said. "The way he's limping around, he's gonna need my crutches soon."

"Maybe I should meet you guys in Dallas," Max said. "I can catch a flight."

"No," Hank grunted. "You'll do more good keepin' an eye on

what's goin' on in Drum. I have no idea what Bingham's gonna do now that he knows Bart killed his little brother. All I can say is it's a good thing Drummond's in a reinforced jail."

"Tell me if there's anything I can do from here," Max said. "Anything."

"Maybe try to call your brother," I said. "See if you can convince him to change his mind and let Carly go."

"I've called and texted him countless times since yesterday, and he hasn't answered."

"Keep trying anyway," I begged.

"Yeah. I will." I knew he would, but I could tell he didn't expect anything to come from it. I didn't expect it either.

"Get some sleep," I said. "I wish I could be there to help you."

"Back at ya, brother," Max said, his voice full of emotion.

I drew a breath. "From here on out, if anyone in law enforcement wants to talk to you, have a lawyer present."

"I haven't needed a lawyer up until this point," he protested. "Won't it make me look guilty if I request one now?"

"No. With everything that's going on with your father and your brother, not to mention your mother taking off, no one will be surprised. Not to mention this mess with Bingham...just humor me, okay? Better safe than sorry. And make sure you get someone with no connection to your family. If the money laundering comes out, you need to be protected."

"Aren't you obligated to turn me in?" he asked with a bitter laugh.

"I plan on resigning from the sheriff's department as soon as I get a new phone."

"Marco," Max said in alarm. "Don't do that because of me."

"It's not just you," I said. "My leg's shot, and it's not that either. I've done..." I stopped and started again. "Let's just say that I'll do whatever it takes to get Carly back safe and sound. Max," I hesitated. "That might mean killing your brother."

Max was silent for several seconds. "Saving Carly is the most important thing. After what he did to Ruth..." His voice broke. "Do whatever you need to do. Wyatt made his choices. Now he has to deal with the consequences."

I was glad he felt that way, even if he'd feel differently if it actually came to that. Because the next time I saw Wyatt, I wouldn't hesitate to introduce him to a few consequences of my own.

CHAPTER FIFTEEN

CARLY

"Why the fuck are we going to Arkansas?" Wyatt demanded, both of us standing behind Marco's Explorer on the gravel road in Nowhere, Tennessee.

I reached up to shut the back hatch. "I'm starving. Get me a decent breakfast, and I'll tell you."

"Don't forget I'm the one with the gun," he growled.

I looked up at him with lifted brows. "You just swore on your life that you wouldn't let anything happen to me. Threatening to shoot me isn't going to work so well now."

He cursed under his breath, then said, "Okay. We'll get breakfast, then you tell me what you have in mind."

Rather than answer, I closed the hatch and started for the front passenger door.

"On second thought," he said. "Sit in the back."

"Why?"

"Because your boyfriend is sure to have put out an APB out for the both of us and his car. It's going to be bad enough driving in his Explorer. I don't need you in the front seat making us look even more suspicious."

I'd considered it too but hoped he hadn't figured it out. I climbed into the back, and he walked up to the side of the door and peered in.

"What are you doing?" I asked.

"Checking the tint. It'll do for the time being, but we're going to have to ditch this car soon." He walked around the front of the car and got into the driver's seat, then backed up the SUV the fifty feet or so toward the main road.

It wasn't even six yet, so the county road was pretty deserted, just trees and crops lining the shoulders.

"Do you even know where we are?" I asked.

He started to boot up the map, then cursed. "Fuck."

I sat up straighter. "What?"

He shook his head and stared at the screen. "Never mind."

Was someone following us? I looked around, but all I saw was open road.

After he studied the map for several seconds, he turned his attention to the road. "There's a town up ahead. We'll grab some breakfast and take it to a motel room to eat."

"Why?" I asked suspiciously.

"Because it makes sense to discuss our plan in private, not some restaurant where someone can eavesdrop."

"Are we planning to spend the night there?"

"I don't know," he said, "but it's worth spending the forty bucks or so it'll cost to get some privacy."

"Forty bucks?"

"We're not staying at a Hilton, *Caroline*," he sneered.

I nearly told him not to call me that, but it would have been wasted breath. I'd already asked him not to countless times. Besides, staying somewhere could work in my favor. The minute he went into the bathroom, I'd either call 911 or run out the door.

We arrived in town and one of the first businesses to come into view was a crappy looking motel. Wyatt pulled off the road and parked in front of the office. After he turned off the engine, he looked over his shoulder at me, indecision in his eyes. Finally, he said, "I need you to cooperate with me. I want to bring you inside, but you can't insinuate that you're with me against your will."

"Why would I do that?" I asked innocently.

Wyatt rolled his eyes. "I'm not stupid. But neither are you. While you've been researching my dad, I've been researching yours. I have information I'm sure you want. Stuff that never made it into the file. If you run off, you'll never get it. But if you stick around, I'll tell you in the room."

I called bullshit. "Why didn't you tell me that earlier?"

"I didn't want to play all my cards," he said with a short laugh. When I didn't answer, he added, "First I'll tell you something intriguing about your former fiancé, Jake."

"Why would I care about that asshole?" I could protest all I wanted, but the wound of his betrayal was still fresh. He'd been my friend for years before he became my fiancé. The cocky look in Wyatt's eyes proved I'd given myself away.

His brow rose suggestively. "I know how you can get even with him."

"I'm not interested in revenge, Wyatt," I snapped.

"Aren't you? Why else would you have gone to so much trouble to bring down my father? Why else would you want to go after *your* father?"

I lifted my chin in defiance. "Justice."

"Just another word for revenge," he scoffed. "Are you saying you plan to just let your ex-fiancé off the hook?"

"I have no way of proving he was involved in all of this," I protested.

A sparkle filled his eyes. "And what if you did?"

Was he bluffing?

"We can work together on this, Carly," he said insistently. "I have information. Information that can help us both."

"If that were true, you could have just told me instead of kidnapping me at gunpoint."

"Okay," he said in exasperation. "I was frustrated and pissed, but I never would have hurt you. I only wanted to punish you. And after I calmed down, I realized you and I can work together. You can be an asset."

"Why didn't you realize this last year?" I asked haughtily.

He groaned. "I've already told you." He pushed out a breath, then asked, "Do you want to know what I know about your father and the dick weasel who used you, or do you want to run back to Marco? Because I suspect I have a hell of a lot more on the both of them than you do."

"Which you claim you're going to hand over to the FBI," I pointed out. "Which means he's going to prison. Which means it doesn't matter what I do or *don't* do. I get the same results."

He gave me a dubious look. "You seriously want me to just hand him over to the feds without you being part of it?"

"All I care about is that he gets what he deserves," I lied, because I *did* want to be part of it. More than I'd realized.

"Liar," he said good-naturedly. "I plan to have you with me anyway, but you get to choose which role you play. Would you rather be the passive princess, or would you like to actively participate in his downfall?"

Dammit. I *did* want to be part of bringing him down. Randall Blakely had always thought his daughter was a spineless weakling. Something deep inside of me wanted to show him and that asshole Jake how utterly wrong they'd both been.

But seek my revenge with Wyatt Drummond? Was I really that stupid?

I couldn't trust him.

At. All.

Still…

I was stuck in this situation for the time being, so I might as well make him think I was cooperative.

"Fine," I said. "I'll go along with your plan for now."

"Good," he said with a satisfied grin. "I knew you'd see things my way, but on second thought, I want you to stay in the car. Just in case our faces are plastered all over the news."

Before I could answer, he got out of the SUV and locked me inside, which was a joke. I could easily open the door and take off running.

So why wasn't I? Why was I sitting in the back seat, watching him walk into the motel lobby?

Wyatt claimed to have information on my father and Jake. I had nothing besides speculation and anger. If he actually had hard proof...

He could be lying, but it wouldn't hurt to stick around a bit longer and find out for myself.

Besides, I had my own Plan B—Skeeter Malcolm—even if it was a long shot. I realized now that I'd been too hasty in sharing my plan to go to Arkansas. At least I hadn't told him why I wanted to go. I'd have to figure out a lie, find a way to escape, and go see Skeeter Malcolm on my own. After taking another stab at getting information from Wyatt.

A few minutes later, Wyatt walked out with a large plastic diamond-shaped keychain with a gold key attached. He slid into the driver's seat and started the engine.

"He rented you a room at six-thirty in the morning?" I asked in disbelief.

"It's empty, so why not? Besides, I only rented it for a few hours. Long enough to get cleaned up and come up with a plan." He put the car in reverse and backed out of the spot. "We're around back, out of view."

"What about breakfast?"

He held up a pamphlet. "The clerk swears by this place and says we can call in an order. There's a phone in the room."

He drove around the building to a narrow parking lot with spaces in front of the rooms. The back of an abandoned-looking strip mall abutted the lot. Wyatt pulled into a space front of a unit in the middle and turned off the engine. "Let's bring in the stuff in the back and go through it to see what we have."

"I know what's in my bag," I said. "I packed it."

"We'll look anyway. Besides, we both need showers," he said, grabbing both guns on the passenger floorboard, then getting out and walking around to the back. I opened the back door and followed him. While I wasn't very fresh, I didn't think showering with him in the adjoining room was a good idea.

After tucking the guns under his left arm, he grabbed both my bag and Marco's, slinging them over his shoulder, then walked to the door with a faded black 14 on the door. After inserting the key in the lock and pushing the door open, he waited for me to enter and then shut the door behind us.

It was a shitty looking room with a single queen-sized bed. The bedspread was in jewel tones and actually had holes in the polyester. The floor was covered in matted green carpet and the TV was chained to the wall. A cheap, tilted desk chair on wheels and no arm rests sat in front of a desk attached to the dresser.

"Nice place," I murmured, keeping my hands close to my body. The room looked like a giant cesspool of germs.

"It serves its purpose." He set the bags down on the bed and handed the flyer from the restaurant to me. "Looks like they serve a typical breakfast. Take a look and tell me what you want. I'll call it in."

I took the flyer. "Coffee," I said, handing it back to him. "I need massive amounts of coffee, along with pancakes, bacon, and scrambled eggs."

He took the flyer with a smug grin. "Hypocrite. You won't let Hank eat any of that."

"Shows what you know," I said, shooting daggers at him. "His blood sugar's been doing a whole lot better lately. Enough so that he gets cheat meals." Okay, so I'd only *just* started letting Hank cheat, but Wyatt didn't need to know that.

How was Hank handling all of this? Would he blame himself? Was he taking care of himself?

Wyatt held up his hands. "Okay. You're right. I've made some bad choices, but I hope to make up for them, and you're at the top of the making-amends list, Carly."

"You're laying it on a little thick, Wyatt."

He gave me an expressionless stare. "Maybe, but I feel like I have a second chance with you, and I don't want to blow it."

I held up a hand. "Let's get one thing straight: I'm with Marco. I *love* Marco. If you think you have a shot at—"

"No," he said, turning serious. "I blew any chances of that happening by stonewalling you for so long. If I could go back and change things, I would." His face fell. "I would change *so* many things. But I can't. I can only go forward and do my best to convince you that I'm on the up and up about wanting to bring your father down."

"For a reward," I added bitterly.

"Do you want me to give up the reward?" he asked. "If that's what it takes to convince you that I'm telling you the truth about my plans, I'll do it."

I groaned and rubbed my eyes. I didn't believe that for a second, but I hadn't gotten much sleep the night before, and I didn't have the mental capacity to deal with his bullshit. "Let's get down to business," I said, crossing my arms over my chest. "In exchange for my cooperation, you promised to give me information about Jake as soon as we got to this room. Spill it."

"Fine," he said, holding my gaze. "The FBI investigated Jake and found something. They plan to indict him next week."

My mouth dropped open. "On what charges?"

"Insider trading."

"With what company?"

"Not your father's," he scoffed.

"Of course not my father's," I said, rolling my eyes. "Blakely Oil isn't a publicly traded company, so whose was it?"

"Some small pharmaceutical company. The tip came from a friend of his from high school. Someone who's in finance."

"Brent Stone?"

"Yeah," he said with a nod, then pointed his finger at me. "That's the one."

I shook my head. "That doesn't make sense. I know Brent. They weren't even friends in school."

"You don't have to be friends to take advantage of an opportunity, Carly," he said with a hint of condescension.

"How did they even get connected enough for Jake to get the information from him?"

"Your ex isn't the nicest guy out there, but he can be charming," he said. "He's fooled a lot of people." His brow lifted ever so slightly. "Including you, if memory serves."

I scowled. "How do you know this?"

"I told you. I've been working with the FBI. This is my last chance to come through for them, and sometimes they tell me things to point me in the right direction."

I narrowed my eyes in confusion. "I don't understand how that can help bring down my father."

"Look," he said in frustration, "I have no idea why they told me that part. To string me along like they have for the last few years? Maybe it's a test to see if I use it to my advantage."

I considered it for a moment. "Seems to me you'd make a killing if you warned Jake it was coming. He's cowardly enough to pay you for the information, then head to some extradition-free country."

"I'm after more than a low-level payout, which is all Jakey-boy could likely pay. Guys like him are leveraged out financially. I'm after my father's land and money, as well as the reward the FBI's gonna give me."

I propped my hands on my hips. "How does knowing about Jake's indictment help me with my father?"

"Carly, you have to be more creative," he said, rolling his eyes as he leaned over to sort through Marco's bag. "If Jake knows the feds are coming after him, he might be willing to rat out your dad."

I shook my head. "So? He'd do that on his own without help from either of us."

"Jesus, Carly," he said in frustration as he looked up at me. "Think bigger. You tell him you have information that will save his skin—and in exchange, you need him to tell you whatever secrets he knows. The key is you'll really have to sell it."

I stared at him in disbelief.

"He won't talk to me," Wyatt said, returning to sorting through the bag. "Why would he? He has no idea who I am. But he *will* tell you. If you handle it right."

I narrowed my eyes. "Your story keeps changing. You said you needed me as bait for my father."

"Yeah," he said in disgust. "With *Jake*. I need you to get him to spill what he knows and then you can confront your father and get it on tape to hand over to the FBI." Tossing Marco's bag aside, he looked up at me. "This way you get to confront them both, Carly. That's got to be worth something."

I hated to admit that Wyatt had a point. If Jake had information that could help me make my father pay, then I needed to take full advantage of it. But I didn't need Wyatt for that. Marco could help me. Except...I didn't want him to compromise his morals, his sense of right and wrong. If he helped me with this, it would change him, and I loved him for being the man he was.

This seemed like a job better suited for Hank, but he was back in Tennessee, and now that this ball was rolling, it was only going to pick up steam. First, I'd have to contact him, then he'd need to get out to Texas. Would I have time to wait?

There was no doubt he'd cross lines Marco wouldn't even toe.

I imagined the satisfaction I'd feel confronting Jake, showing him that I wasn't the scared woman he'd tricked for years. And the look on my father's face when he realized *I* was going to be his downfall.

Before, I'd wanted justice for my mother's murder and my father's plans to kill me, but now…now I wanted to make them squirm.

Maybe I deserved to be with Wyatt after all.

CHAPTER SIXTEEN

CARLY

"Happy you stuck with me now?" Wyatt asked with a smirk.

I gave him a dry look. "I'm not ready to commit one way or the other yet."

He grinned, then glanced at the door to the parking area before turning back to me. "Why don't you take that shower now?"

"I don't want a shower. I want breakfast."

He lifted a shoulder. "I haven't even ordered breakfast yet, and then I have to go get it. Take a shower while I'm gone."

He planned to leave me alone while he picked up the food? I wasn't opposed, more like surprised. I could use my alone time to call Marco. Sure, he didn't have his cell phone since Wyatt had tossed it out the window last night, but I had no doubt he'd get one as soon as he could. And when he turned his new phone on, he'd have a message from me. I'd rather talk to him and make sure he was okay—I was worried about his leg—but I need to assure him that I was okay and would be in touch again as soon as possible.

Wyatt grabbed the TV remote from the dresser and began flipping through channels until he found a station reporting the Nashville morning news. He studied the screen for a few moments, then set the remote on the dresser.

"Looking for reports about us?" I asked.

"I need to know how much heat we have on our tail."

"And if we have a lot of heat?" I asked.

He grabbed his black bag and pulled out the laptop he'd stolen from Marco's kitchen table. Would he become furious again when he saw all the files Hank and I had uploaded?

"What's the password?" he asked once the screen came to life.

I considered telling him I didn't know, but I figured this was a pretty harmless way to prove I was cooperating. "LettyandSmoky," I said. "The L and S are uppercase. No spaces."

He tapped on the keyboard and looked surprised when the computer woke up. But his bad mood returned a half minute later when he realized the cheap motel didn't have internet.

He tossed the laptop onto the bed and turned back to the TV with a worried look. The possibility of us being in the news had to be a real concern for him. He was driving Marco's stolen car, which would be easy to spot. Especially if someone saw us together.

"I thought you were going to order our food," I said, dragging his attention from the screen.

"I have to see if it's safe to go out." He shook his head, then grumbled under his breath, "I'd kill for internet."

"Guess you get what you pay for," I deadpanned.

"Sorry it's not the Ritz," he muttered as he walked over to the nightstand.

"I don't need the Ritz," I said. "A Hampton Inn would do just fine. What happened to your phone?"

"I left it in the van during the commotion," he said in a low growl.

He picked up the phone on the bedside table and called the number on the pamphlet. After placing our order, he gave them the name Jack.

When he hung up, he said, "They said it'll be ready in forty-five minutes, so I'm using the time to take a shower."

I was uncomfortable with the idea, but I didn't want to call too much attention to my unease, so I didn't say anything.

At least until he added, "You're going to have to come into the bathroom with me."

"Excuse me?" I protested, then shook my head vigorously and pointed to the bathroom door. "There's no fucking way I'm going in there with you."

Surprisingly, he gave me a sympathetic look. "I realize we've reached a truce, but I still don't trust you not to run off. So it's either you sit in there with me, or I tie you up. Your choice." Then he added, "If I tie you up, it won't be like last night. I'll just make sure you're not going anywhere."

I frowned. "Then tie me up, because I'm not seeing you naked."

He laughed, then slid his belt out of his jeans. "Put your hands behind your back."

I fought the urge to resist, because I knew it wouldn't get me anywhere, and put my hands behind me. He wove the belt around my wrists a couple of times, then said, "Go sit in that chair." He pointed to the beat-up office chair sitting at the built-in desk.

Panic started to race through my blood, but I forced myself to calm down. He was just going to tie me to the chair. It wouldn't be like the restraints he'd put on me in the car.

There was a support bar attaching the top of the back support to the chair base. I sat down and he threaded the belt around the bar and then fastened it on the outside. He gave it a tug to test it. Then, content, he moved in front of me. "What do you want to watch on TV?"

"How long of a shower do you plan on taking?" I asked sarcastically.

Laughing, he grabbed the toiletry bag out of Marco's travel bag and headed into the bathroom. The shower turned on, and I could hear him moving the shower curtain rings across the bar as he got in, but he left the door open.

Asshole.

I kept my gaze on the television and the perky morning news anchors. They were talking about some summer country music festival in Nashville, but I wasn't paying much attention. Instead, I was questioning my decision to not make a run for it when Wyatt had gone into the motel office.

But I'd been cooperative when he'd tied me up, which must have earned me some trust points. Giving him the computer password too. If he trusted me, I'd be able to duck him, and then…

And then what?

Was I just deceiving myself because I suspected Wyatt's idea of a confrontation with my father would be a whole lot more sordid than what Marco had in mind?

What did it say about me that I might prefer Wyatt's plan?

Think long and hard about this, Carly…

Wyatt began to hum in the shower, probably using Marco's shampoo, and I reminded myself that no matter how Wyatt tried to dress it up and stick a pretty bow on it—he'd kidnapped me at gunpoint. He was holding me against my will. He claimed to be working with the FBI, but I found it difficult to believe they'd condone kidnapping.

I still wasn't sure I believed him about any of it, but I was stumped on how he would know that information about Jake. Granted, I hadn't Googled Jake's name in a month or two, but I hadn't seen his name in the news other than in his portrayal of a distraught fiancé.

Several commercials played on the TV screen, and then the newscasters were back, informing viewers of local news. Police were investigating a murder west of the city. There'd been a pedestrian hit and run downtown, and the victim had suffered serious injuries. Finally, they reported a disturbance east of Crossville with multiple gunshots off an exit ramp on I-40. While state police were investigating, no other information was being released.

"Huh," Wyatt said from behind me. "So they're not telling people to look out for us. Maybe the FBI is coming through for me after all."

I'd been so engrossed in the news report I hadn't heard the water turn off.

Would the FBI really condone him kidnapping me? But I had to admit I couldn't see why they *hadn't* mentioned our names. "I guess that means we're free to go about our business," I said, confused over my torn feelings.

I wanted Wyatt to get caught, right? But I couldn't help thinking he might know something else that was useful given what he'd shared about Jake. But that would mean sticking with him for the time being.

"How do you plan to get out of the kidnapping charges?" I asked, not looking behind me. I had to admit part of me was worried that he was naked.

"For one thing," he said, moving closer, "you're the one who would have to bring the charges against me, and I'm hoping you'll see this worked out in your favor."

I pursed my lips and kept quiet. I highly doubted he could get out of this scot-free. I wasn't the only one involved. Wyatt had witnesses who likely saw him holding me at gunpoint. He'd shot at Marco and injured his leg. I highly doubted the authorities would let all that go. Wyatt was no fool. He had to know it too.

He walked past me with a skimpy white towel wrapped low around his hips, the top corners gathered in his hand. His back was to me as he dug around in Marco's bag.

I turned away, keeping my focus on the TV. Out of the corner of my eye, I saw the towel drop to the floor. "For fuck's sake, Wyatt. Do you really have to do that out here?"

"I've got nothing to be ashamed of," he said.

I could tell there was movement to my side, and I hoped to God he was putting on his clothes—scratch that—*Marco's* clothes. The thought of him even touching them, let alone wearing them felt viscerally wrong. Like they should be burned after Wyatt discarded them wrong. And now he was parading around naked like it was the most natural thing in the world. Another surge of anger that Wyatt had put me in this situation mushroomed in my chest. Why hadn't he just told me the truth last year when I'd begged him to?

I was watching a traffic report when he moved in front of me. I started to squint my eyes closed until I realized he was wearing another of Marco's shirts and a pair of his jeans.

"I'm going to go get breakfast now."

I held his gaze, not liking what he was hinting at. "So untie me before you go."

He chuckled. "Carly, if I don't trust you while I'm in the bathroom, I'm definitely not going to trust you while I'm gone."

I jerked my hands, shaking the chair. "Untie me, Wyatt!"

"Patience," he said with a laugh, digging his wallet and Marco's key out of the pocket of his old jeans. "The sooner I leave, the sooner I'll be back with food."

"Then lock me in the bathroom," I said more eagerly than I'd intended. "I can take a shower while you're gone. You're the one who made that offer earlier."

"Good try, but that wouldn't work out in your favor or mine. The lock on the bathroom door is on the inside, and even if I *did* figure out a way to lock you in, there's no window for you to escape." Chuckling again, he walked to the door. He started to open it, then turned back and said, "Oh, and feel free to scream all you want. I told the clerk we're into rough sex and the only other person here is a deaf guy at the end. See you in a bit."

He walked out and shut the door.

I fought the urge to scream anyway. The asshole had just gotten the better of me, and I felt like an utter fool.

Why had I considered staying with him?

Take a deep breath. This is better than the situation in the van. I gave myself a moment for the panic to ease, then forced myself to start working on a way out of this.

The belt buckle was somewhere on the outside of the chair support, but if I twisted the leather, maybe I could move it to the inside and unfasten it.

How far away was the restaurant? How much time did I have?

Likely not enough. And if Wyatt came back and found me mid-escape, it would ruin all the groundwork I'd laid. So I made myself watch the irritatingly chipper news anchors, forcing myself to be patient.

But the news show ended, and Wyatt still didn't return. How long had he been gone? A half-hour? He'd said the breakfast place was close. He should have been back by now.

The national morning show came on, and another half-hour later, he still wasn't back. He'd been gone at least an hour. Had something happened? Had he been caught?

How long would it be before someone found me?

Calm down.

He'd only rented the room for a few hours. But as filthy as the place was, how long would it be before someone came in to clean? It wasn't like the place was bursting at the seams with customers.

Another fifteen minutes passed before I decided I couldn't wait any longer. Even if the police found me, which I wasn't so sure about, I might still be in trouble. There was a good chance my father would get alerted. He had enough power and money to insist the authorities bring me straight to him. They probably would, too, since I was a documented missing person.

I couldn't let that happen.

It was time to act.

I began to twist the belt, making slow work of sliding the strap to move the buckle inside the chair. After about ten minutes, I could feel the edge of the buckle, but it still wasn't close enough to unfasten.

I continued tugging for another five minutes or so, then nearly cried when I finally felt the cool metal of the buckle edge touch the tips of my fingers. But it was too soon to rejoice just yet. I still couldn't unbuckle the belt.

It took me another five minutes before I could touch the prong, and I started to wonder what I'd do once I got myself free. Call Marco, obviously, but it was unlikely he'd gotten a phone yet. I didn't have any money or ID, so I couldn't rent a car or get a bus ticket. Maybe I could find a coffee shop and wait it out until Marco could come get me.

One thing was certain: I couldn't call the police. I couldn't trust them not to cave to my father's demands. The sad truth was, if I called Marco at this point, he'd feel compelled to call them, if for no other reason than the badge he carried. And if he didn't call them at my request, he'd be forced to live with the regret.

I'd moved the belt enough that I could finally touch the prong and had just started to try to pry it loose from the belt when I heard movement outside the door.

I froze, my heart pounding against my chest. Was it Wyatt? Was it the police? I was screwed either way.

If it was the police, they'd ship me right back to the man who'd murder me.

If it was Wyatt, I'd be back to square one with getting him to trust me. He'd realize that I'd been manipulating the belt.

The door slowly pushed open, and I held my breath, telling myself the police would have barged in or at least made their presence known outside the door. Then my mind leapt to a new possibility. What if Wyatt had enemies who had caught up to him and wanted to use me too? I was sure Louise wasn't too happy with him, and she was crazy enough to stalk him and then haul me to Texas and hand me over to my father for the reward.

Wyatt's face appeared in the opening, wearing a big grin as he said, "Honey, I'm home."

I hung my head as relief flooded me, which filled me with shame. I shouldn't be happy to see him. He'd kidnapped me at gunpoint.

He'd hurt Marco and gotten Jerry killed. He'd betrayed Hank and used his grandson for his own personal gain.

Wyatt Drummond was a bad person, and instead of revulsion, I felt relief when he walked through the door. I choked back a sob.

What the hell was wrong with me?

But it was easier to redirect that self-loathing on to him. "Where the hell have you been?"

He walked through the rest of the way carrying a paper bag and a drink tray with two large cups of coffee. "I was getting breakfast."

"Where the hell was the restaurant?" I demanded. "In Arkansas?"

"I had a few errands to run." Setting the tray on the dresser next to the TV, he gave me a perplexed look. It was then I noticed he had a couple of shopping bags slung over his arm. "You seem upset."

"I thought you got caught, Wyatt!" I shouted.

His brows shot up. "And this bothered you?" he asked sarcastically. "If anything, it seems like you'd be relieved."

"I had no idea if you'd tell the police where I was. I had no idea how long I'd be stuck here until someone found me."

He dropped the food bag on the bed. "I told you I only rented the room for a few hours. Someone would have found you this afternoon."

"Look at this place!" I cried out, nodding toward the bed. "It's a fucking dump! I wouldn't be surprised if no one came to clean the room for days! And even if the police *did* come find me, they'd probably ship me off to my father." I ground my teeth. "And that is *not* how I plan to go back to him."

He knelt in front of me, concern on his face as he lifted a hand and smoothed back my hair from my damp cheeks. "It's okay, Carly. I'm not going to let that happen."

"Don't leave me like this again," I said, then another surge of anger rushed through me, and I jerked on the belt. "Either we do this together or not at all."

He looked into my eyes and then nodded, saying softly, "Okay. We do it together."

The way he looked at me, his gentle touch, and his soft words unlocked something in me, and I realized I *wanted* to trust him. I *wanted* to believe we really were in this fight to bring down my father together.

Don't be stupid, Carly. He's using you. He's always used you. This wasn't the time to start believing him. But I still needed him to think I did.

"I didn't think you were coming back," I said, genuine tears falling down my face. If I couldn't sell this, I was in big trouble. The last thing I wanted was for him to hogtie me in the Explorer. "I panicked, Wyatt, so for the last fifteen minutes, I've been working on getting the belt to get loose." I started crying again, forcing it a bit. "I couldn't let the police find me. I wasn't trying to get away from you."

His eyes darkened. "But you *were* trying to get free."

"I told you, I've only been working on it a short time. Maybe fifteen to twenty minutes. If I'd *really* wanted to get away from you, I would have been long gone by now."

He spun the chair so my back was to him, my heart thumping crazily in my chest. Then, several seconds later, he spun me back around to face him. He gave me a cold, hard stare.

"Aren't you going to untie me?"

He remained expressionless. "Seems like you can do it yourself. You were so close."

My anger was back. "Fuck you, Wyatt."

A dark smirk crossed his face. "If I remember correctly, you told me that was off the table." He rose and moved over to the bed.

"You're really going to leave me tied up?"

He sat down and reached into one of the shopping bags. "My father always insisted that if Max or I started a job, we needed to finish it ourselves." He waved toward me. "Be my guest."

"I'm not sure using your father as a role model is a good life strategy," I sneered.

He shrugged, then pulled a long, thin package from the bag.

"You'll probably be relieved to see I got a new toothbrush." He waved it around as proof. "I'll even let you have this one since you were so offended that I used yours."

I made a look of disgust, but he laughed and pulled out a larger package. "I also got a new phone." He turned it around to show me a pay-as-you-go phone, then set it on the bed. "It's not as fancy as an iPhone, but we're both used to no cell service, so I think we'll be okay."

He took one of the coffees out of the tray and took a sip. "Aww, just what I need to help me perk up."

I gave him a hard stare. "Are you really going to leave me tied up?"

"You can get yourself free anytime you like, Carly." He drew in a deep breath. "That bacon sure does smell good."

My stomach audibly growled. *Dammit!*

I started fumbling with the belt prong as Wyatt pulled a Styrofoam carton out of the food bag and lifted the lid. The smell of pancakes and bacon wafted into the air, making my stomach growl again. Pulling a plastic fork out of the bag, he sawed off a piece of pancake and popped it into his mouth. "Mmm."

I got a finger wedged under the prong as I used my other hand to give the belt strap some slack.

He picked up a piece of bacon and took a bite, then leaned closer to me and waved it in front of my face. "Need a little extra motivation?"

I shot daggers at him while I strained to get the prong free from the hole in the leather.

"Want a bite?" he asked, holding it up to my mouth.

I didn't answer, instead, I got the prong free and then slid the strap through the buckle, loosening the binding on my wrists.

He sat back on the bed as I pulled my arms free and shook them out in front of me. "Do you have a thing for BDSM you never told me about during our three weeks together?" I asked sarcastically.

"As quickly as you got yourself free, I could ask you the same," he said with a grin, reaching into the paper bag and pulling out a second Styrofoam carton and setting it on the bed closer to me.

I started to protest, then realized what he was suggesting. "So you believe me that I just started working it loose?"

He shrugged.

I considered staying in the chair, but my butt was sore, so I got up and grabbed the container, opening it while I stood. Snatching up a piece of bacon, I took two bites.

Wyatt reached his hand behind him and slipped a pocketknife out of his back pocket. He flicked it open as he stared at me, as though searching for a reaction.

Was that supposed to be an intimidation tactic? It wasn't lost on me that he hadn't pulled a pocketknife out of his jeans earlier to transfer to Marco's pants. Marco also didn't usually keep one in the Explorer. He must have bought this while he was out.

He broke eye contact, then picked up the plastic package encasing the burner phone and started to open it with the knife. After he got the phone free, he plugged it in, giving me a mischievous look.

"For fuck's sake, Wyatt. For once in your life, quit playing games. What are you up to now?"

He laughed. "Who says I'm up to something? Maybe I'm just proud of you for getting yourself free."

"Yeah, right," I grumbled as I opened the syrup container, then picked up a pancake and dipped it in. Wyatt hadn't given me a fork, and I was too hungry for table manners. "Forgive me if I don't believe you're encouraging me to find a way to escape."

"Contrary to what you think, Carly, I don't want you broken. If you're going to help me face your father, then I need you strong."

"Aren't you worried I'll try to escape?"

He shrugged as he picked up his food container. "You were right. If you really wanted to escape, you would have been long gone before I came back. I believe that you tried to get free because you were worried the police would show up."

I stared at him in disbelief.

"Look," he said off-handedly. "I figure we need to start proving we can trust each other."

"Plus, you need me docile in the car since the windows of Marco's Explorer aren't well tinted."

He shrugged again, scooping some eggs onto a plastic fork. "Not to worry. I've taken care of that situation."

"What does that mean?" But I already knew he must have gotten something else to drive, which only made sense. The police were looking for Marco's Explorer. It explained why Wyatt had been gone so long. "What kind of vehicle did you get?"

His mouth twisted with amusement. "A van. Gotta keep you out of sight." He paused for another bite of bacon, then continued, "Like you said, your father knows you're alive and well, and we can't have him finding you before we reach Texas." He gave me a pointed look. "You were saying something about needing to go to Arkansas earlier?"

A tight smile stretched my lips. "Only because I'd love to stop by and say hello to my old friends."

"Yeah," he said, leaning back. "I don't think that was it."

"What else could it be?" I asked innocently.

His eyes darkened. "Do you want to know what Jake will tell us about your father?"

"What do you think?" I shot back.

"Then maybe you should try bein' straight with me, and I'll do the same," he said. "I gave you the information I had on your ex, so now it's your turn."

"And what if I don't have anything to tell you?"

His eyes turned hard. "Then I'd say you're lying."

"Look," I said, dipping my pancake in syrup. "You're bringing me with you one way or the other." I jutted my head forward. "You've already admitted that you need me."

"And I'm trying to build a trusting partnership here, Carly," he said, his face flushing with frustration.

I didn't answer for a moment, then said, "Look, it's not easy to forget that less than twenty-four hours ago, you were holding a gun to my head, okay? How do I know you're not going to snap and actually kill me in an hour or two?"

"Because I need you for my plan." Scowling, he picked up his Styrofoam tray and set it on his lap. He pulled out a small package of syrup, opened it, then dumped it over the pancakes. "I'm gonna be straight with you," he said earnestly. "I know more incriminating stuff about Jake, stuff that will convince him to talk, but I'm not going to tell you anything else I know until we get to Dallas." When I started to protest, he held up a hand. "Are you really that surprised? It's my insurance plan. If you want to know what I know, you'll have to stick with me." He studied me for a moment, then took a bite of his food. When he swallowed, he asked nonchalantly, "So what was your plan once you got loose?"

"Honestly?" I shrugged, letting my guard down a little. "I wasn't sure. I figured you hadn't come back because something had happened to you, but I don't have any money to rent a car or get a bus ticket."

"You weren't gonna call Marco?" he asked in surprise.

"Oh, I was going to call him. I need him to know I'm okay."

"Hadn't figured out where to run?" he asked skeptically.

"I wasn't sure I *wanted* to run."

He released a short laugh. "You really expect me to believe that?"

"Whether you believe it or not is up to you, but it's true. I'm torn right now, and I need more time to sort it out. You were right that I want revenge. And as much as I hate to admit it, I really want to know what you have on Jake." I had to sell this, so I added, "Plus, I like that you're a little more heartless than Marco. You'll make sure Randall Blakely suffers." I pushed out a breath of frustration. "I'm not sure what that says about me, and honestly, I'm struggling with it."

There was a grain of truth in it. I loved Marco, and I loved that he had a line in the sand separating good and bad. I used to have that line too, but my father didn't. Neither did Wyatt. It was going to take someone who wasn't afraid to get dirty to bring my father down. But I also couldn't trust him, which meant I couldn't rely on him.

I'd stick with him, find out what I could, then I'd ditch him and figure out where to go from there. I had no doubt Marco was headed to Dallas, and I could reconnect with him there. Because doing this without Marco's help was foolish. I needed his moral compass to help me see this through.

Right?

Wyatt sobered, then said softly, "That makes you human, Carly."

I wasn't so sure about that. I was pretty sure it made me a monster.

Shared DNA or not, it made me more like my father.

Like I said, monster.

CHAPTER SEVENTEEN

CARLY

Wyatt picked up the other coffee container and held it out to me. "We're both gonna need our caffeine today."

I reached for it and started to take off the lid.

"It's already got creamer added."

I took a sip, surprised it was doctored to the taste I preferred. I took another drink, also surprised the coffee was actually hot. Which meant he'd run his other errands first, then grabbed our breakfast—though that made sense, on second thought. If he'd stolen a van, it wouldn't have been smart to transfer our breakfast while doing it.

Don't read too much into the doctored coffee and hot breakfast, I warned myself. Up until this point, Wyatt had *never* had my best interest in mind, and he wasn't about to start now. He was only trying to make me more complacent.

I set the coffee down on the nightstand and picked up a piece of bacon. "So your plan," I said thoughtfully, "is to go to Dallas and arrange a meeting with Jake?"

He grabbed his container and stabbed some eggs. "It seems like a good one. Got something better in mind? Something to do with

Arkansas?" He waggled his eyebrows.

I considered my options. Going to Jake might not be such a bad idea, but was it enough? Besides, we were hinging our entire plan on the belief that Jake would talk. There was no guarantee that he would. I needed a way to get information about my father on my own, and I knew a potential way to get it. Or at least *try* to get it.

I suspected if I arranged a meeting with Skeeter in prison, Wyatt wouldn't be able to go in with me. Which meant he wouldn't know what—if anything—I'd found out.

Even better.

"My friend Rose," I said as I ate another piece of bacon, "I think she can help." Besides, Rose was married to the Fenton County chief deputy sheriff. Joe would help me out of this situation with Wyatt should I decide I needed an out.

"The woman who took you in when you arrived in Arkansas?" he asked in surprise.

I kept my gaze on my food, sneaking glances at him. "Not her directly, but I need her to set up the meeting. Without her, I doubt I'll be able to see *him*."

He turned his full attention on me. "See *who?*"

I lifted my gaze, watching for any sign of recognition. "Skeeter Malcolm."

"Skeeter Malcolm?" Confusion flashed in his eyes as though the name sounded familiar but he couldn't place it.

James "Skeeter" Malcolm's name had been in the news, but most people wouldn't have paid any attention. Wyatt wasn't most people, though, and I didn't trust his supposed lack of recognition. Especially if he'd been looking into my father the way he'd claimed to be.

"He had a connection to my father, and he's currently in a federal prison awaiting his trial. He made a deal with the FBI to provide information that would bring the Hardshaw Group down, but he didn't show up at the actual raid. It still took place, and the FBI got what they needed, but they reneged on their end of the bargain because he didn't show. They still took down the Hardshaw Group, but the FBI—"

"Were a bunch of dicks," Wyatt filled in with a derisive grunt. "Surprise, surprise. But what makes you think he'll agree to see you? And why would he tell *you* anything?"

"Because of Rose," I said. "She worked with him in the underworld, and he respects her. In fact, he wasn't at the raid because he was trying to help her save her kidnapped baby."

His gaze narrowed. "Seems like more than just a working relationship."

I gave him a long, hard stare, because he didn't need to know a single thing more about Rose than what I was telling him. Wyatt Drummond was the last person I trusted with their secrets. "He owed her. Skeeter Malcolm can be a real bastard, but he pays his debts, and he owed one to Rose."

"And you think he'll agree to see you if Rose sets it up?"

"I have no guarantee of that, but it's worth a try."

He pressed his lips together and seemed to consider it. "How do I know this isn't some kind of trick?"

"You don't, just like I don't know what you're going to do next, but like you said, we have to try to start trusting each other or we'll never get this mission accomplished." I held his gaze. "This is me trying to prove you can trust me."

He fell silent and ate several more bites before he finally said, "So if I agreed to this plan, how would you go about it?"

"I'll need to contact Rose, of course. I'll need to call her."

His brow lifted. "Call? Not see her in person?"

Marco and I had been planning to go there in person, but if I took Wyatt to her, I could potentially be putting her and her baby in danger. No matter how big of an asshole Wyatt was, I didn't want to believe he'd hurt a baby. Still, I wasn't willing to take the risk.

"She lives in Henryetta, which is in southern Arkansas. Skeeter is incarcerated closer to the Tennessee border. It would save us time to go straight to the prison."

He studied me as though trying to find a hint of duplicity. "Do you think you can get it set up today?"

"I can try," I said. "I have no idea how it all works. Or if Skeeter will even agree."

He nodded. "Do you know her number?"

"I don't have her cell phone number, but I can call her plant nursery, and if she's not there, I'll tell them I need to talk to her. They'll give me her number."

"Why would they do that?" he asked skeptically.

"Because I used to work there when I lived with her. They know me."

He considered it for a moment, then lifted his brow. "You know the number for the nursery, but you don't know her cell number?"

I shrugged slightly. "The nursery's number is a play on their name. The letters for Gardner match the numbers on the keypad, but they don't open until nine." I cast a glance at the digital clock on the nightstand. "That's another half hour."

"So let's say you get this appointment with Malcolm, do you think he'll actually tell you anything useful?"

"I don't know," I admitted, "but it's something, right? It's worth a try. It's on the way to Texas anyway."

"You want to visit a federal prison when there's a good chance the police are looking for the both of us?" Wyatt asked in disbelief. "Seems like a good way for me to get caught."

"*I* could get caught," I pointed out. "It's a risk for me too. They could send me to my father. Consider this, in the story about us on TV, they said the Tennessee State Police are looking into the shootout off the highway. There was no mention of either of us. No photos or names. The shootout was in Tennessee. We're going to Arkansas, not to mention a *federal* prison. Marco says those agencies are notorious for not talking to each other. We'll be fine."

"Don't be naïve, Carly. They're looking even if our names aren't on the news."

"Okay, maybe so. But it can't hurt for me to call Rose," I said insistently. "Because if this isn't even possible, there's no point in talking about it."

He turned to look at the TV. The morning news show was talking about a group of kids who'd run a lemonade stand to raise money for cancer research. Wyatt looked totally enthralled, but I knew he was thinking about my suggestion. At least he hadn't thrown it out.

"*If* you call her," he finally said, still not looking at me, "it will be on speaker, and I'll be listening to every word. If I don't like the direction of the call, I'm ending it immediately."

"Okay. I can live with that."

A deep scowl crossed his face. "You realize that meeting with this guy is a long shot. So long of a shot, it would be a stupid bet."

"Then I can ask Skeeter about the odds when we see him. He used to be a bookie."

We finished our breakfast in silence while the TV played in the background.

"Shouldn't we be leaving town?" I asked. "Didn't you just steal a car?"

"We're good."

"Care to elaborate?" I asked with a pointed stare.

He sighed. "I don't think the van's going to be missed for a while, and I figure we might as well stay here until after you make your call. We'll take off after you're done."

We watched TV, both of us wound up with anxiety, and I questioned whether we should go ahead and leave to puncture that feeling. But I kept the suggestion to myself. Let Wyatt think he was calling the shots right now, and I'd call them later.

At 8:57, he picked up his burner phone. "Okay. Let's do this. Gardner, you say? Area code?"

"501."

He punched in the numbers and the phone rang over speaker. My stomach twisted into knots. I hadn't spoken to any of my friends in Arkansas since I'd left last October. It had been a painful, teary goodbye, only days after Rose's sister's death. When I'd shown up in Drum, I'd been mourning her passing. Wyatt was the first person I'd met. It felt strangely poignant that he'd been there in the beginning, when I'd arrived in Drum, running from my father, and now he was with me at the end.

I suspected that made me a sentimental idiot.

"Gardner Sisters Nursery," an older woman answered in a cheerful tone.

Tears flooded my eyes. "Maeve?" I asked, my voice breaking.

"Yes," she responded, sounding hesitant. "Who's this?"

"Carly."

She paused so long I started to think she'd forgotten me.

"*Carly?*" she asked in a gasp. "Where are you? Are you okay? You left, and I never got to say goodbye."

Very few people knew the truth about where I'd gone—Rose and Joe, and Neely Kate and her husband, Jed, who was the one who'd gotten my new identity and car. They'd known secrecy was paramount, though, and they would have kept that information to themselves.

"I know. I'm sorry. It was all very sudden."

"Are you okay? You don't sound okay."

My heart filled with love. Maeve was a woman in her late fifties or early sixties who mothered everyone. She'd definitely mothered me.

"I'm okay, and I'm hoping to come visit soon, but right now, I really need to speak to Rose. Is she around?"

"She and Baby Hope are walking through the greenhouse with Neely Kate and Baby Daisy. Let me take the phone out to her."

"Thank you."

I could hear Maeve's breath as she walked toward the greenhouse. "How have you been?" she asked. "Where did you go?"

I considered lying, but I didn't see the point anymore. If this foolhardy plan to get my father didn't work, I couldn't go back to Drum. "Eastern Tennessee. I've been waitressing in a tavern out here. I found a place that took me in like y'all did. And I'm with a wonderful man, Marco." Tears stung my eyes again. "I can't wait for you to meet him."

Wyatt sent me a hard glare.

"Aw…" she said softly. "I'm happy for you, Carly." Still, she seemed uncertain, as though she didn't trust that I was telling the full truth. Which was fair since I wasn't entirely.

"Thanks," I said. "And I can't wait to catch you all up when I come visit. Hopefully, that will be soon."

"I can't wait for that," she gushed. "You've been greatly missed, and I'm sure Rose and Neely Kate would hire you back in a heartbeat if you decide you want to move back home."

"Thanks, Maeve," I said past the lump in my throat. "That means the world to me."

Maeve's voice grew fainter as I heard her call out to Rose.

Wyatt pressed mute. "Was it really necessary to give your life story?"

"I knew this woman for months," I said in exasperation. "It would have been rude to not tell her about my life, not to mention suspicious."

"Carly?" I heard a familiar voice cry out into the phone.

"Just watch yourself," Wyatt growled as I hit the unmute button.

"Rose?" I asked, close to crying again.

"It really is you!' she exclaimed, her voice breaking. "Is it safe for you to be calling?"

"If you're asking if my father has been taken care of, the answer is no. Not yet. But that's why I'm calling."

"Should I get Joe?" she asked, sounding concerned.

"No," I said. "This seems more like a Jed kind of situation."

"Oh."

"Carly?" I heard Neely Kate say. "Hey, girl, it's me."

"I miss you all so much," I said. "I owe you and Jed everything—"

"You hush your mouth," she teased. "We always help our friends. What do you need now?"

I took a breath and looked up at Wyatt, still wondering if this was a good idea, but I was committed now. "I need to talk to Skeeter."

Neither one of them said anything, and I checked the screen to make sure the call hadn't been disconnected.

Finally, Neely Kate said, "Why in the bejeesus would you want to do *that*?"

"Because I know he has information on my father."

"Carly," Rose said sympathetically. "He's not going to tell you anything. If he was going to talk, he would have told the FBI. If he knows something, he's decided to keep it to himself."

"I don't plan to take it to the FBI," I said. "I plan to use it to get my father."

"What does that mean?" Rose asked, the words laced with concern.

"I'm going to confront him."

"Alone?" she asked in alarm.

"No." I looked over at Wyatt. "Definitely not alone."

CHAPTER EIGHTEEN

CARLY

"Who are you planning to do this with?" Rose demanded.

"A friend," I said, having to spit out the word.

Rose picked up on it. "What kind of *friend*?"

"It doesn't matter," I said, gripping the phone tighter. "Rose. *Please.* My father knows I'm alive, and he's getting close to finding me. The only way to make sure I'm truly safe is to put him away."

She hesitated. "Carly, I don't know. I'm not sure James will even agree to see you, let alone help. And he definitely won't do it for me."

"He'll do it for Jed," Neely Kate said.

"Neely Kate!" Rose protested.

"Carly's right," Neely Kate told her. "If her father is close to finding her, then she's not safe. We were hoping to neutralize him with the big bust, which didn't happen. We need to help her come up with something else."

"She can't face her father with a questionable *friend.* She needs to be with someone she trusts. She needs to be with *us,*" Rose said. "Carly, come to Henryetta."

Wyatt shook his head and mouthed, *No.*

"I can't," I said, even though I wondered if I should try to ditch

Wyatt and do just that. But I couldn't let on that I was considering it. I needed to make Wyatt trust me. "I'm fine, Rose. *Really*. I need to do this alone." Then I quickly added, "With my friend."

Both women were silent for several seconds.

"Let me call Jed," Neely Kate said. "Give me your number so I can call you back."

I didn't have the number for the burner. I glanced up at Wyatt, who had, other than mouthing his refusal to go to Henryetta, remained silent during the call.

He gave me a dark look, then grabbed a piece of paper from the bed and handed it to me. It was the paperwork that had come with the phone, and it had the number on it.

"You ready?" I asked.

"Yeah."

I gave her the number and she read it back.

"I'm going to call Jed now," Neely Kate said, "then I'll call you right back. When are you looking to do this? Next week?"

"Today."

"*Today?*" she and Rose both said in alarm.

"My father knows I'm going by Carly Moore and that I'm in eastern Tennessee," I said. "It's only a matter of days before he finds me."

I heard a baby coo in the background and Rose make a hushing noise.

"I'm not sure they'll be able to get you in today," Neely Kate finally said. "They have to do a background check and—"

"It has to be today," I said. "Tomorrow at the latest. My father is closing in." And so were the state police. Once they caught up to us, I'd lose my opportunity to see Skeeter. Not only would Skeeter refuse to see me, but law enforcement would be suspicious of my request.

Something else occurred to me: there was no way Marco could be involved with any of this. If Marco wanted any kind of career in law enforcement, then he couldn't be any part of meeting a man in

Arkansas who had worked with one of the head members of an international crime syndicate.

The cold hard truth was sinking in: Marco couldn't be involved with this anymore.

Oh God. Did that mean I really did have to do this with Wyatt?

If the FBI had enough evidence to arrest my father, I'd wholeheartedly let them take over, but I wanted to be there when they arrested him. But if not...

My testimony that my father planned to have me killed wouldn't do any good. It would be his word against mine, and I knew how that would end—with me dead.

I was done hiding. I was taking back control of my life.

If Wyatt could help me with that goal...

I knew he was using me, but that didn't mean I couldn't use him too.

"I have to see Skeeter," I said, unashamed that I was begging. "I need this to be done."

"Okay," Rose said softly. "I won't stand in your way. In fact, tell me what else I can do to help."

"If you have any way of convincing Skeeter to talk to me, I'm begging you to do it. But if not, maybe Jed can try."

Both women were silent, then Neely Kate said in a weary tone, "I'll do my best, but I can't make any promises. Still. If anyone can make this happen, it's Jed."

"Thank you," I said, my voice cracking with relief. "I appreciate it more than you know, Neely Kate."

"Hang tight." Then she hung up before I could say goodbye.

Wyatt stared at me with a hooded expression. "I don't like it."

"Don't like what?"

"All of it. We need to forget this and head straight to Texas. The state police are on our tail, and for all we know, your friends know that and are setting us up."

"They would never set me up," I said, "but I wouldn't be surprised if one or more of them shows up at the prison when I go to see Skeeter."

He shook his head. "No. No fucking way."

I lifted my chin. "If you want me to help you lock in your deal with the feds by dealing with my father, then you'll agree to this." I held his gaze. "And one more thing. I'm calling Marco."

He burst out laughing. "You really think I'm going to let you call him?"

My voice hardened. "I know you will, because you need me, and if I'm cooperative, it makes your life a whole lot easier."

His laughter died down, and his glare returned. "If you think I'm letting him join us, you're a fool."

"Even if I wanted him here, the two of you together are like oil and water. Especially after you *attacked* him."

"I didn't want to hurt him, Carly," he said, his anger fading, "but I also couldn't let him take you. I could have shot him, you know."

We'd already had this discussion, and I didn't see any point in rehashing it.

"Look," I said, my voice softening slightly. "I know he's worried sick about me. I need to let him know I'm okay. And I need to make sure he's okay. His leg is really messed up from when Carson Purdy shot him. I'm worried about him."

He sobered. "I wasn't even thinking about his leg when I swung at him, Carly. All I knew was that I needed to get away from him, and I didn't want to shoot him. I planned to just knock him down then take off and leave." He searched my face. "It's important that you believe me."

I wasn't sure I did. Hitting Marco in his injured leg to distract him seemed like exactly the sort of underhanded bullshit Wyatt would pull. But I saw no reason to tell him so. "It was a stressful situation, and we're lucky no one was killed. But I'm sure you can see why I need to call him."

"So you can tell him where to find you," he said with a sneer.

"That would be stupid since we're about to leave."

"You could tell him about your plan to see Skeeter Malcolm."

"But then you wouldn't take me there, so there'd be no point." I took a jagged breath. I really, *really* needed to sell this. "Marco's too by the book, and if I'm going to work with anyone who doesn't follow the rules, it should be you. Besides, you're ultimately doing this for the FBI, right? So it's a win for both of us."

He looked dubious.

"But just because I've decided to do this with you doesn't mean I'm leaving Marco completely in the dark. He's worried about me, and he needs to know I'm okay. Besides, you tossed his phone out the window and the stores aren't open yet, so I'll just be leaving him a message."

He started to say something then stopped.

"If we're working together, then we have to trust each other," I said in a firmer tone. "No more tying me up. No more treating me like a prisoner. This is the first test for me to prove that I see us as partners."

His brow lifted. "Partners?"

"Partners. I'm not happy about it, I can assure you that," I insisted, curling my upper lip. "I'd rather have a root canal, followed up by a pap smear, but this is the best way to accomplish what I need, so I'll suck it up and deal with it."

"Way to sugarcoat it."

I held his gaze. "I'm done sugarcoating shit to please men. So deal with it."

His lips tipped up with the barest movement.

"So I'm calling Marco and telling him I'm okay. That I have a plan and will check in with him later. He doesn't need to know it's with you. In fact, he'll feel better if he doesn't hear your name at all."

Wyatt got up and walked to the window, shifting the curtain slightly to the side to surveil the parking lot. "Fine. Make your call, but we're not waiting long to hear from your friends. If they don't come through, we'll stick to my plan."

I didn't answer, just picked up the phone and punched in Marco's number. It went straight to voicemail and hearing his voice message brought tears to my eyes.

The beep sounded and I said, "Marco. It's me. I'm okay. Don't worry about me. I have a plan, and I'm going to follow through with it. I'll keep you updated as I go, but for now, just know that I'm okay." I paused, needing him to know this part in case he figured out I was still with Wyatt. "Also, Wyatt didn't kill Ruth. He just told me that to scare me. In fact, he was really upset when he told me that he was there when Franklin killed her."

I drew a breath, struggling with the lump that had formed in my throat. "I love you. I love you *so much,* Marco. You've stood by me like no one ever has, and you have no idea how much that means to me. Even though we've lived through hell, I've been happier with you than I've ever been in my entire life." I wiped a tear that trailed down my cheek. "I'm going to fix this so we can spend the rest of our lives together with no one looking over our shoulders. No more living in fear. But you have to trust me." Then I hung up, close to tears. I wasn't fool enough to think he'd let this go, but I hoped he'd at least know I was okay.

Wyatt was still by the window, looking through the crack. "He really makes you happy, huh?"

"He's honest with me, even if I don't want to hear what he has to say, and after all the dishonesty in my life, I need that."

He turned to look at me, letting the curtain drop. "I never stood a chance."

"Maybe you would have if you hadn't kept so many secrets," I said with a weary smile. "But I don't think you can help it. If it wasn't secrets about your father, it would have been something else. Maybe it's part of your Drummond DNA or you learned it at your father's knee, but duplicity is part of who you are."

He studied the floor for a few seconds, then ran a hand over his head. "Yeah," he finally said, "maybe you're right."

"But you can change," I added because my words seemed to

break something in him. "This mission with me is a good place to start."

He hesitated, then nodded before looking up. "Do you really plan to do this with me?"

"Yeah," I said, my voice breaking. "I do."

Why did it feel like I was starting to believe it?

He took a step toward me, then stopped when the cell phone rang. His eyes hardened again, and he picked it up from the bed, checking out the number. "It's a 501 area code."

I stood and took the phone from him. "It's Jed."

"How do you know?"

I couldn't pull it out of thin air before, but seeing it now, I knew.

"I just do," I said, then answered and put the call on speaker. "Hey, Jed."

"Carly," he said in what sounded like disbelief. "How are you doin', girl?"

"I've been better, but I'm good. I hope Rose has been passing on my messages in our emails."

"She has, but I'll feel better when I see you face to face."

"That's gonna have to wait," I said. "Until after I get this done."

"And you hope to accomplish this by visiting Skeeter?" He sounded skeptical.

"He knows something about my father, otherwise he wouldn't have had a deal with the FBI."

"He's not going to tell you what he knows, Carly. He won't even tell me."

"What can it hurt to try?" I pleaded. "Even if he gives me the smallest morsel of information, it's better than nothing."

He heaved out a sigh. "You're not going to let this go, are you?"

"No, my father caught onto my trail again."

"Damn, Carly. I'm sorry."

"I'm tired of sitting around waiting for him to find me. I'm going after him, and I think Skeeter Malcolm might have something useful to help me."

He was silent for several seconds, then said, "He'll want something in return."

"Like what?"

"Information. Something sent to him. Maybe checking on someone. I don't know, but he won't do it for free. The only person he'd give a free pass to was Rose, and she thinks that bridge has been burned."

"*She* thinks?" I questioned. "You don't?"

"No. That bridge will never be burned, but she's too racked with guilt to believe otherwise." He drew a breath. "All I'm saying is be prepared for him to ask you for something."

"I'm good with that," I said resolutely. "Just make this meeting happen."

He hesitated as though still debating with himself, then said, "I think the best way to make this happen expediently is through his attorney. Carter Hale has a few wheels greased on Skeeter's behalf, and he can visit Skeeter and say you're an assistant or something."

"Rose and Neely Kate said I'll need a background check. That might prove problematic."

"You have a criminal background I don't know about?" he said with a chuckle.

"No, but I was in the middle of a shooting last night, and I know the Tennessee State Police were notified. Marco would have told them my name is Carly Moore." I'd gathered that much from the lack of a news update on the missing oil heiress Caroline Blakely. "But that means they'll be looking for Carly since they think she was kidnapped."

"*Have* you been kidnapped?" he asked in a deadly tone, reminding me that he used to be Skeeter Malcolm's right-hand man for a reason.

"It's a long story, but I'm doing this of my own free will."

"So Carly Moore's out? What about Caroline Blakely?"

I hesitated.

"While Caroline is a missing person," Jed said, "as far as law enforcement is concerned, no foul play was necessarily involved. Last I checked, they think you were a runaway bride."

"I don't have Caroline's ID. Won't I need that to get in?"

"I've got it," he said quietly. "I've held onto it. I'll have Carter check to see what kind of flags your real identity will set off if you use it."

"Okay." Somehow it felt weird to consider going there under that name. Like I would be putting on a facade. I may have been born as Caroline Blakely, but I'd spent most of my life feeling like an imposter. Carly Moore fit me. When this was all over, I was having my name legally changed.

"I'm gonna call Hale now," Jed said. "See if we can get something set up for this afternoon. I already called his office. He's in court now, but he should be out within the next hour. I'll let you know as soon as I hear something."

"Thank you, Jed. You have no idea how much I appreciate this."

"Last year, I told you I'd take care of you, and I don't make vows like that lightly," Jed said. He paused, then added, "Skeeter's at the federal prison in Forrest City, Arkansas. How far away are you? You can't be that far if you want this to happen today."

Since Marco and I had considered doing this ourselves, I'd looked up the location of the prison, and I knew the general area of where we were now. "I'm guessing four hours," I said. "Maybe five."

"Carter's gonna need time to get this arranged, not to mention it's gonna take him about four hours to get there himself, so we're looking at late afternoon. Does that work?"

"Yeah," I said. "That works."

"It's good to hear from you, Carly," Jed said. "We'll get this sorted out."

I only hoped he was right.

CHAPTER NINETEEN

CARLY

I hung up and looked up at Wyatt. His face was a mask, which told me he didn't like something about my phone call. Maybe all of it.

"What?" I demanded.

He snorted. "You think I'm letting you go inside that prison without me?"

"I'm sorry. It can't be helped."

"You're not doing this," he said, propping his hands on his hips. "No fucking way."

I got up and stood in front of him. "I told you that you have to start trusting me, and this is how you do it. I'm bringing something to the table with this meeting."

He threw his hand toward the phone. "*Hero Jed* just said he doesn't think Skeeter Malcolm is going to tell you anything."

I ignored the dig about Jed. "And *then* he said Skeeter would expect something in return."

"Just like my dad's favors," he spat in disgust.

He was right, so there was no arguing that. "It can't be helped."

A derisive laugh bubbled out of his chest. "I'm sure that's what all those poor suckers who got ensnared in my father's orchestrations said too."

I groaned in frustration. Maybe I *was* just like those poor, desperate fools.

"I'm going to take a shower," I said, turning my back to him and grabbing my bag. "I can't be stinking like sweat from your perverted kidnapper van when I go into that prison."

"I'm not a pervert!" he protested as I went into the bathroom.

I set my bag on the closed toilet lid and turned to face him. "No, you're just guilty of aggravated kidnapping. That's *so* much better." I shut the door and locked it.

Let him stew on that.

I spent fifteen minutes in the shower, washing my hair, shaving my legs, and then rehashing my decision to stay with Wyatt. But in the end, I decided I needed to stick with him until it felt like I couldn't get anything else from him. Maybe I used him to take me to the prison and figured out what to do from there.

I finished my shower, then put on a sundress, the closest thing I had to something a legal professional might wear to visit a prisoner.

When I got out, Wyatt was on the bed watching some morning talk show. He cast a glance toward me, surprise in his eyes.

"I have to look like his assistant," I said dryly. "And in all honesty, we should probably stop somewhere to get a suit jacket for me to wear over it."

He snorted. "What a great idea. Let's go on a shopping trip."

I rolled my eyes and turned to face the mirror. Using the motel hair dryer—I was shocked they had one—I dried and styled my hair, then put on makeup.

Wyatt kept stealing glances at me from the bed, but I ignored him.

By the time I finished, Jed still hadn't called. I really didn't want to sit here waiting with Wyatt, and I didn't need a fourth coat of mascara. Besides, I wasn't entirely sure where we were or how long it would take us to get there.

I closed the tube and tossed it into my makeup bag. "I think we should take off and head toward Forrest City."

"You're presuming your hero can make this happen," Wyatt shot back, his gaze on the TV. Apparently, his jealousy wasn't limited to Marco.

"He will." I put my makeup case in my bag and dug out a pair of sandals. I'd need to get a pair of pumps when I got the jacket.

I scanned the room, making sure I hadn't forgotten anything. "I'm ready. Let's go."

Wyatt shot me a darkly amused look. "You think you're suddenly calling the shots?"

"If you want me to cooperate with Jake—which I really think you do—you'll go along with me on this." I headed for the door. "Now let's go."

To my surprise, he did as I said, shoving a few things into his bag and picking up the phone off the bed.

I stood on the sidewalk outside the room, staring at the red minivan parked in front of our unit. It looked like it was a decade old. "You traded Marco's Explorer for this hunk of shit?"

"Wow," he said sarcastically. "What happened to Nice Carly? Remember her?"

"You lost Nice Carly when you pointed a gun at my head."

I walked to the passenger side and opened the back door, then tossed my bag onto the bench seat and climbed in after it. I needed to get to Forrest City, which meant I didn't need Wyatt getting pulled over because someone recognized me through the windshield.

Wyatt cast me a questioning glance, but he didn't protest. Instead, he got behind the steering wheel, twisting between the seats to put his bag in the back

"Do you have a map of Arkansas?" I asked.

"Do I *look* like I have a map of Arkansas?" he sniped as he turned on the engine. "The burner phone's too basic to have maps."

"Well, we're going to need a map to figure out how to get to

Forrest City." Then I added with plenty of snark, "Unless you already know how to get there. I know you've been in prison, but I thought it was state, not federal."

He shot me a look that suggested I wasn't as funny as I thought, then turned to face the steering wheel. He backed out of the spot and headed toward the road.

"I figure we'll run south of Nashville then catch I-40," he said once he'd turned onto the four-lane highway running through town. "We have to go through Memphis to head to Dallas anyway."

"But we'll need to get a map to verify that," I pushed.

"Jesus Christ, Carly," he growled. "Enough with the map shit."

He was understandably on edge. He was a wanted man, and he was out in the open again. It wasn't lost on me that he'd stolen a van without a navigation system. He was probably worried the police could use the GPS to track us. I'd noticed his reaction to the system in Marco's Explorer. In fact, I couldn't help wondering why they hadn't found us sooner. "Where'd you get the van?"

He kept his gaze on the road, his jaw tight. "You don't want to know."

I suspected he was right. The less I knew, the better. Could I be considered an accessory if they figured out that I'd had an opportunity to escape and hadn't taken it?

I'd worry about that later.

Wyatt switched on the radio and a decades-old country song filled the van. He continued through town and headed back out into the countryside. I nearly told him to change the channel, but it wasn't worth the fight.

About a half-hour into our drive, the cell phone rang. Wyatt picked it up and snuck a glance at the screen. "501."

"It's probably Jed again. Who else would be calling?"

"Marco. You called him from this phone, so it's only a matter of time before he tries to call you."

Why hadn't I considered that? I was going to blame lack of sleep and coffee. But then again, why would Marco have a phone with an Arkansas area code? "Can he track that phone?"

"Yep," he said, smacking the p.

"Then why did you let me call him?"

He kept his gaze on the road. "To prove that you can trust me too."

My mouth dropped open, but he handed me the phone and I quickly answered before it went to voicemail. "Jed?"

"Sorry to disappoint," a man said in a snide tone, "but this is Carter Hale."

I gasped. Skeeter's attorney. "Mr. Hale. Thank you for calling me."

"Don't go thanking me yet. Jed says you want to see my favorite client. *Why?*"

"I already told Jed that I'm hoping—"

He cut me off, sounding bored. "Yeah, I got all that. I would advise my client not to divulge any information that we still might be able to use to help him with his situation. Or undermine our ongoing negotiations. Information is power, Ms. Blakely."

So Jed had also told him who I was. I decided to try to use it to my advantage. "It works both ways, Mr. Hale," I said in a firm tone. "I might have information I can share with him."

"Mr. Malcolm doesn't give a shit how much cream your father puts in his coffee or how stiff he likes his collars starched. You have nothing of interest to him."

I rested my forehead in my hand. He was right. I had nothing other than what I'd overheard the night of my rehearsal dinner, but then again, that wasn't nothing. I sat back up and drew in a breath. "I *do* have information. Why do you think I ran away?"

"Because you overheard your father and your fiancé discussing your impending demise," Carter Hale said dryly. "He already knows, and it doesn't benefit him in any way."

"But I overheard them talking about something else. Something that might be to Mr. Malcolm's benefit."

"Doubtful."

Shit. Part of me wanted to ask if Jed was there, but I was pretty sure dealing with Carter Hale was like playing patty cake compared to Skeeter Malcolm. If I couldn't finesse this conversation, I didn't have a chance in hell in succeeding.

"Are you sure?" I challenged. "I overheard them talking about my father's business—the illegal one. I plan on confronting my father face to face so I can see his reaction when I tell him he's going away for a long, long time. But I need more information to make that happen. If Mr. Malcolm is interested in getting something particular from him, I can do my best to get it."

He laughed. "Do your best?"

"It's all I can promise," I said, "but I did just help take down a tyrant in eastern Tennessee who was responsible for several murders and ran the whole damn town, so I can get shit done."

He chuckled. "Yeah, right."

Dammit. Now what was I going to do?

"You need to pull out your inner bitch," Wyatt said, low enough that Carter Hale couldn't hear him.

He was right.

I squared my shoulders and dug deep. "And what if I have something that Mr. Malcolm actually wants? Or if there's something he wants me to get for him? How will he like it when he finds out that he had an opportunity, and you threw it away?"

Carter hissed, then swore like a sailor, stringing together several colorful phrases before he said, "Goddamn you, Caroline Blakely."

"I'm pretty sure I'm already damned, but you can add it to the tab," I said stiffly. "Now are you going to set up this meeting or do I start composing a chatty letter to Mr. Malcolm? Do you spell Hale with an A-I-L or A-L-E?"

"I'll call you back," he grunted, then hung up.

"What's your gut on this?" Wyatt asked, glancing at me in the mirror.

I sat back in the seat and stared out the side window. "He'll set it up." I turned back to face the front. "Why did you help me?"

He glanced at the road, then back up into the mirror. "Like I said. Proving you can trust me."

I highly doubted that was possible, and yet...wasn't that exactly what I was doing?

CHAPTER TWENTY

MARCO

A few hours outside of Nashville, Hank pulled over at a rest stop. I struggled to get any rest. My leg was throbbing. I leaned the seat back, trying to stretch my legs out, but I was too tall for the space. On top of that, I was worried about Carly. I kept seeing the look on her face when she recognized my car off that exit ramp. She'd looked so hopeful, and I'd let her down.

I tried to push her out of my mind. I needed rest if I was going to be any good to her. But that didn't help any, because my mind skipped straight to Rodney Bingham's body, buried outside Max's parents' home all these years.

Todd Bingham was sure to be on the warpath. Bart was in jail. Emily was in hiding. And Wyatt had stolen Carly. Would he take his frustrations out on the only Drummond left in Drum? Max was too young to have had anything to do with it, but—

Jesus, had Wyatt known?

There were no answers, and I eventually drifted off into a troubled sleep.

We roused around six and used the restroom at the rest stop. Wyatt had my bag, so I didn't have my toothbrush or a change of clothes, but Hank freshened up with the things he'd brought.

When we left, we found a McDonald's for breakfast. I couldn't help thinking how upset Carly would be that I'd let Hank eat a McGriddle and hash browns, but I didn't say anything as we pulled through the drive-through and got back on the highway.

"I figure we should get to Memphis about the time the phone store opens," Hank said. "We can get new phones there."

"Yeah," I said, shooting him a questioning glance as he wolfed down his sandwich. "Should you be eating that?"

"No," he scoffed, "but what Carly don't know, don't hurt her."

"Uh, I think it actually hurts *you*. Do you have your insulin?"

"I got some in the back. Don't you be worryin' about me," he said gruffly.

"Aren't you supposed to be checking your blood sugar and givin' yourself shots?"

"You think I'm just gonna do that in the car so some passerby thinks I'm some druggie? I did it in the bathroom."

"Okay." I hadn't seen him take anything in with him other than his duffel bag, and I was pretty sure his insulin had to be refrigerated. But there was a cooler in the back floorboard, so maybe I'd just missed it. Hank had run a major pot ring. The man was obviously good at subterfuge.

Leaning my elbow against the window, I rubbed my throbbing temple. "Are we doing the right thing by going straight to Dallas? Should we stop and think this through more?"

"We know Wyatt plans to take her to Texas. It only makes sense to head there."

"Yeah, you're right." And he *was* right, even if it felt wrong.

"I say we go check out Blakely's office and home." He shot me a look. "You know where he lives?"

A lump filled my throat. "I got the address from Carly." We'd planned to surveil it together. What if we got there too late? What if Wyatt had driven straight through to Texas? What if he turned Carly over before we could get there to stop him?

"Good," Hank said with plenty of attitude, "we'll scope it out and maybe try to find the bodyguard who worked for Blakely."

I nodded. "Carly told me about him right before she was kidnapped. Duncan Morris. He was part of her father's security team. Carly said he was fond of her when she was a kid."

"That's him. The asshole who provided half her DNA said Morris quit about a year after Carly's momma died." His brow lifted. "Sounds suspicious as hell to me."

"You think something happened to him?"

"Or he felt guilty." His hands tightened on the steering wheel.

"Maybe he was part of creating Carly's mother's accident."

"Or he knew about it and quit. The DNA donor thought Morris was alive, but I'd be surprised if Blakely let someone with that kind of evidence live off the payroll." He paused. "I sure as hell wouldn't."

I cringed. "I really don't want to hear that, Hank." I wasn't stupid. I knew he'd done questionable things, but that didn't mean I wanted details.

"No sense sugarcoatin' the truth," he said gruffly.

"So we need to find out if this guy is even still alive," I said. "In addition to scoping out Randall Blakely and finding Carly." I rubbed my eyes, feeling overwhelmed and desperate.

"She'll find a way to reach out to us," Hank said with a firm nod. "I know it in my gut. She's a smart girl. Resourceful. Plus, deep down, Wyatt likes her. He won't hurt her, and she'll find a way to use his feelings for her to her advantage."

"You're making her sound manipulative," I said, my anger rising. "She's not like that."

"She's not the girl who showed up in town last November," Hank said. "And from what she's told me, she's not even close to the woman she was when she ran away. She's grown a backbone, and she's not gonna roll over and play dead for Wyatt. She's gonna find a way to best him."

"You think she's gonna sleep with him?" I asked in disgust.

"Fuck no," Hank said with a sneer. "No pun intended. She won't sell herself like that, but she might play along with him to gain his trust." He turned to look at me. "But we need to replace our phones ASAP. Then we wait for her call."

Several hours later, we were standing outside a cell phone store in Memphis, waiting for the doors to open. A half-hour later, we had our new phones.

Our information was in the process of being transferred from the cloud, and I watched the screen as notifications populated it.

"There's a voicemail from a number I don't recognize," I said excitedly.

"It could be the state police with an update," Hank said.

My heart sank. He was right, but I remained hopeful as I opened the voicemail app and played the message. I nearly collapsed when I heard her say, "Marco, it's me."

Hank's eyes widened.

I nodded. "It's her."

"What's she saying?"

I stopped the message and set it back to the beginning, then moved away from the counter and the curious salesperson. Hank followed. I set the phone to speaker and turned the volume down before pressing play. The same relief as before rushed through my blood at the sound of her voice, but then it turned to ice as the message played.

"Marco, it's me," she said, her voice soft as a caress. "I'm okay. Don't worry about me. I have a plan, and I'm going to follow through with it. I'll keep you updated as I go, but for now, just know that I'm okay." She paused. "Also, Wyatt didn't kill Ruth. He just told me that to scare me. In fact, he was really upset when he told me that he was there when Franklin killed her." She paused again, and when she spoke again, her voice was tight. "I love you. I love you *so much*, Marco. You've stood by me like no one ever has, and you have no idea how much that means to me. Even though we've lived through hell, I've been happier with you than I've ever been in my entire life." She paused again, then she turned insistent. "I'm going to fix this so we can spend the rest of our lives together with no one looking over our shoulders. No more living in fear. But you have to trust me."

That was the end.

I stared at the phone in relief that she was alive, but dismay that she hadn't told me where she was or what she was doing.

"What the fuck does that mean?" I asked Hank, my panic back. "Do you think she got away from Wyatt and wants to do this on her own?"

Hank's gaze dropped to the phone, and he was silent for several seconds before he looked up at me. "No. I think she's still with Wyatt, but she's convinced him they want the same thing."

"She's working with him?" I asked in dismay.

"Or she's letting him think she is," he said. "She doesn't sound like she's in distress, so that's one less worry. Now she seems convinced that Wyatt didn't kill Ruth."

"Do you believe that?" I demanded, furious that she'd found the opportunity to call me, but she didn't tell me where she was.

"I have a hard time believing it, but at this point, I don't know."

I took a step back, outrage filling my entire being. "I'm supposed to be less worried that she's *okay* with that madman?"

Hank's face turned murderous. "Keep your voice down. I'll leave you the fuck behind if you're gonna act like a jealous lover."

"I'm not jealous."

"The fuck you are," he grunted. "But you need to keep your shit together while we get these phones squared away. Then we'll figure out what to do next."

I didn't answer. Hank didn't give me a chance to. He hobbled over to the counter and paid for our things, another thing that stuck in my craw. I could have paid for my own phone, but it would have nearly emptied my bank account. Rural county deputies didn't get paid all that much.

While we waited, I pulled up the Ford app. After multiple attempts with failed passwords, I finally logged in and found the location of my Explorer. It was still in Tennessee, in some little town southeast of Nashville.

"I found my car," I said, shoving the phone at Hank.

He stared at it skeptically. "Carly ain't there."

"She might be."

He shook his head. "Wyatt's no fool. He dumped it."

"But what if she *is* there?"

His mouth twisted to the side. "You've got a few options. One, we tell the state police and let them check it out. Two, we head over there and deal with Wyatt Drummond ourselves—if he's there, which I strongly suspect he's not. Or three, we ignore it and keep heading to Texas."

"You're forgetting option number four," I said. "We call her."

He considered my suggestion for a moment. "Let's do it outside."

I was anxious and unsettled when we went out to the car. Hank headed to the passenger side. "You're driving."

"What about your driving apparatus?" Hank had a special lever for the gas pedal so he could drive with his leg amputation.

"It comes out. Just toss it in the back."

"We're wasting time," I grunted as I got in and glanced down, trying to figure out how to remove it.

"A couple of minutes ain't gonna mean shit in the scheme of things," Hank said with a frown, "and I'm sure as shit not drivin' all the way to Texas."

He was right. It only took me a couple of minutes to remove the lever and toss it into the back seat, then pull out my phone.

"You know she's not gonna answer, right?" he said.

"So you think we shouldn't even try?" I asked in disbelief.

"Of course we should try," he grumped like I was a simpleton. "But it means we need to come up with a message to leave her.

We've got to have a plan." He held my gaze. "And I need to be the one to leave it."

"Why?" I demanded.

"Because Wyatt considers you an annoyance, but he'll damn well take me seriously."

"What the—?"

"You're a lawman," he snapped. "You follow the rules."

"Not anymore," I said through gritted teeth. "Not if it saves Carly."

"Well, Wyatt doesn't know that, now does he?" Hank countered. "He still sees you as a deputy sheriff. But he knows I ruled the area with an iron fist and didn't take shit from anyone. If he hears from me, he'll be running scared, because not only does he need to worry about surviving his encounter with Daddy Blakely, but he'll need to deal with *me*."

"What if he gets so scared, he kills Carly to save himself?"

"He won't," Hank said, full of confidence. "I'm tellin' you, he likes her too much to hurt her."

My anger mushroomed. "Is that supposed to make me feel better?"

"Yes," he snapped. "Of course it is. Stop thinkin' with your dick and use that brain God gave ya."

I drew in a deep breath. He was right. Of course he was right, but I still didn't like the implications.

"I'll leave the message," Hank said, pointing a finger at me. "And you'll sit there and keep your mouth shut."

I didn't respond, but he must have taken my silence as agreement because he pressed send. The phone rang several times, then went to a generated voicemail greeting.

After the beep, Hank held the phone close to his face and said in a warm voice, "Carly, it's Hank. You need to watch your back and take care of yourself, girlie." His face hardened and so did his voice. "Wyatt, I know you're with her, so you listen and listen good. You hurt one fucking hair on her head, and I will kill you in a death so painful, you'll be begging me to end it. You got my grandboy killed, and I'll be damned if you let anything happen to Carly. So if you come to what little fucking sense you have left, call me at this number and we'll work something out. Because if I find out you turned my girl over to that monster who called himself her father, there is no place on this earth you can hide from me. I will *find* you and I will make you *pay* for every last thing you did to me and mine."

He hung up, his eyes burning with rage, and he shot me a look of disgust. "What the fuck are you waiting for? We need to get to Texas. We have work to do."

CHAPTER TWENTY-ONE

CARLY

We hit construction traffic before we joined back up to I-40 in Tennessee, which slowed us down. Carter Hale called back shortly after he'd hung up and told us we had a 3:30 appointment, but we needed to meet him in the parking lot fifteen minutes early.

Around 10:30, Wyatt's phone rang with Marco's number. My heart lodged in my throat when I leaned over the seat and saw who was calling.

"We can't answer it, Carly," Wyatt said firmly, yet there was a touch of sympathy in his words.

I drew in a deep breath. "I know."

"He'll leave a voice message," Wyatt said, his voice softening. "We'll listen as soon as it pops up."

I nodded.

Sure enough, less than a minute later, the message popped up with a voicemail. Only it wasn't from Marco. When Wyatt opened it and put the phone on speaker, Hank's voice boomed from it.

Tears sprang to my eyes as I heard him threaten Wyatt if anything happened to me, but I also saw the pain on Wyatt's face.

When the message ended, I couldn't help asking, "Was any of it real, Wyatt? Or was it all manipulation?" I could have been ugly. I felt used, and I knew it was even worse for Hank, but seeing Wyatt's reaction softened something in me.

He waited so long to answer, I thought he was going to ignore me, but finally he said, "It was real."

"With Hank and Seth," I said to clarify.

"You too." His mouth twisted. "If I could do things differently, I would. But I can't," he said with resignation. "I can't change a damn thing. All I can do is try to fix things going forward."

"By trying to prove you're trustworthy."

His gaze lifted to meet mine in the mirror. "Yeah. And not just because I want to be with you, Carly, although I do. That never stopped. But because I owe it to you to help you bring your father down."

"And to help yourself," I added with a touch of sarcasm.

His eyes twinkled with mischief. "Would you trust me if I said my motives were completely selfless?"

"No."

"See?" he said with a laugh. "Truthful."

I tried to get him to talk about what had happened with Seth, but his face tightened with what looked like grief, and he shook his head. "I can't talk about that. Not yet."

I didn't press. Baby steps.

So instead I asked him about Ruth and Franklin. He repeated his story about their tussle for the gun, his eyes haunted. Franklin had held her in his arms as she died and then run off.

"I think he parked somewhere and killed himself," Wyatt said, looking haunted by the thought.

"You don't think he took off so he wouldn't get arrested?" I asked suspiciously.

"No. He was very upset. He would have accepted prison, only he thought that was too good for him. He said he'd killed the only person he wanted to spend the rest of his life with." His gaze met mine. "He couldn't live with himself."

"And you didn't try to stop him?"

His voice broke. "I was too upset over Ruthie."

"What car did he drive off in? Because Ruth's car was in the driveway, and his truck was in the shop."

Irritation hardened his jaw. "I don't know, Carly. Like I said, I was too busy freaking out over Ruth."

"What did you do with the folder?"

"What folder?"

"Don't play stupid," I said, getting irritated. "The folder full of spreadsheets for your mother." When he didn't answer, I added, "The hard proof that Max was laundering money for your mother?"

Recognition filled his eyes. "I didn't see it, I swear."

"That's convenient."

"It's not like I came there looking for anything," he said, annoyed. "I came there to tell Ruth about her baby sister and walked into the Franklin mess. After Franklin took off and I got my shit together, I left too."

"Why didn't you call the sheriff and report it?"

"What good would it do?" he countered. "She was already dead, and I knew how bad it would look if I called it in. So I left."

"And then accused me of being responsible for her murder."

His lips pressed tight. "In a way you were."

"That's a load of bullshit and we both know it," I snapped.

"Okay," he admitted. "It was bullshit, but I was still trying to get you to drop the whole thing with my father. Your meddling was interfering with my own plans. I needed you to back off."

I still wasn't sure I believed him. My mind flitted to his notes. "Why did your notes in those files sound so calculated?" I asked. "It was like you were an outsider dropped into Drum to snoop on people, not someone who had lived there his whole life."

"I was turning those files over to the FBI," he said. "I couldn't write them like a journal. Dear Diary, I had lunch with Carly, and I love the way she smiles. The way one side of her mouth lifts slightly higher than the other. And the way her eyes sparkle when she's happy. Or the sound of her laugh." He lifted one shoulder in a shrug. "Not exactly something the FBI wants to know, and frankly, I didn't want them to know how much you meant to me. They'd only use it against me. So I painted it to look like you were just a means to an end. Hank too."

I had to admit that his explanation made sense, but I couldn't ignore all the other things he'd done. I was about to ask him about Jerry again, but he beat me to it with a question of his own.

"So what's the deal with you and Todd Bingham?"

I released a bitter laugh. Why was he asking again? "Do you think we're sleeping together?"

"No, although I wondered if you were before. I fully admit I was jealous as hell. Still am."

My chest tightened. I had to shut this down. "Wyatt. There is no you and me. I'm with Marco."

"I know," he said. "And I won't try to get in the way of that, but I'm not going to pretend I don't have feelings for you. Now that we're being truthful."

"Is that what we're doing? Being truthful?"

"That's what *I'm* doing," he said defensively. "You're not?"

"News flash," I said, leaning forward, "I've *always* been truthful with you. You're the one hoarding secrets like a squirrel getting ready for winter."

His brow lifted. "You've always been truthful with me? You've told me *everything*?"

"I did until I couldn't trust you, Wyatt," I countered angrily. "I would have been an idiot to tell you everything then."

We rode in silence for several seconds, me fuming.

"Look," he finally said, "you're right. I'm wrong. We both know I've been a first class asshole."

I crossed my arms and remained silent.

"But I'd really like to know more about your relationship with Bingham."

"I don't have a relationship with Todd Bingham," I snapped.

"You have *something*."

I huffed out a breath. "We have an understanding, or at least we used to. I'd give him information he found helpful, and he'd shared information with me." Then I added, "I think he liked it when I asked him to sell me a car instead of getting one from you. It helped prove I wasn't your..." I struggled to fill in the appropriate word.

"Woman?"

"Yeah, I guess. I know he's trouble with a capital T, and I know he's up to something with all his research into his family tree and the history of the town."

He sat up straighter. "He's still interested in that?"

"He is. I just haven't figured out why." I sat back in my seat. "Max thinks Bingham treats me different not only because I give him information at times, but also because I don't kowtow to him like everyone else does." I tilted my head to the side, and said in an aside, "He also thinks it might get me killed someday."

"He's right. Todd Bingham is a stone-cold killer."

"I know." I'd learned that only a few days after coming to town.

"You were loyal to him," Wyatt said in an accusatory tone, gripping the steering wheel.

"Only if I was in the middle of working out a deal with him. But I admit I was more loyal to him than you most of the time. He actually followed through on his promises. You never did."

"I deserve that," he said.

"I know."

He laughed. "Inner bitch Carly is still out."

"Even if it wasn't too late to put her away, I need her to deal with Skeeter Malcolm."

"And your father."

My back stiffened. "One asshole at a time."

CHAPTER TWENTY-TWO

CARLY

We took minimal bathroom, gas, and food breaks, and reached Forrest City a little after three. Which meant we didn't have time to stop and pick up anything else for me to wear. I only had my sleeveless sundress and sandals. I hoped the prison didn't have a dress code.

Wyatt slowed down as we drove into the prison complex, and I could see he was having second thoughts.

"I'm not ditching you, Wyatt," I said, even though I wasn't sure that was true. I was still playing it by ear.

He kept his eyes on the road. "That's not all I'm worried about."

"Are you worried they'll know who I am when I check in?"

He didn't answer, but then he didn't need to. I was worried about that too.

"We should have asked Carter Hale to see if there were any flags attached to my name," I said.

"Not a good idea. No reason for them to think you're with me under duress."

"But I'll be using Caroline, not Carly."

"Marco surely gave a statement to the state police," Wyatt said. "He may have told them who you really are."

The thought had occurred to me as well. And there was no denying that if Jed suspected Wyatt had kidnapped me, let alone at gunpoint, he would likely smash his face in and then drag me back to Henryetta.

I had to admit it was tempting.

Over the past few hours, I'd done a lot of thinking about my situation and the best way to proceed after our side trip to Forrest City. Jed was a man of many talents and likely better suited than Wyatt to help me take down my father. Between him and his friends, who came from questionable backgrounds, I knew they'd be my best bet. But Jed had a new baby and a wife who adored him. I couldn't do anything to jeopardize his and Neely Kate's happiness. Which meant sticking with Wyatt for now.

Still, the closer we got to our destination, the more my resolve faded.

"Surely they would have been alerted if there was a problem during the background search," I said.

"If there's an alert on your name, it's going to say you were kidnapped. Hero Jed's not exactly going to call you up and ask you about it. He'll address it when we show up." He shook his head. "This is a bad idea, and not just because I'm worried about myself. Even if they don't tie your Carly Moore identity to Caroline Blakely, your father might have told the authorities that you're mentally unstable. If that's the case, they'll take you straight to him and won't let you have a say in the matter."

"Then I'll talk to Jed and get his opinion."

"Hero Jed," he sneered.

"Shut up, Wyatt," I said with a sigh. "He's happily married with a new baby. He has no interest in me other than as a friend."

"Right."

His insinuation pissed me off. "The people in Henryetta treated me better than I'd ever been treated in my life before coming to Drum. I loved them and spending my last month caring for Rose's dying sister only made us all closer. Jed was the one who purchased my new identity and helped me on my way. He's like a brother, and I'm not going to let you cheapen that with disgusting insinuations."

He lifted his fingers off the steering wheel in surrender. "Whatever."

"Your emotional maturity was really stunted, Wyatt. You might want to work on that when we're done with all of this."

He laughed, but then quickly sobered as he pulled into the parking lot.

A dark sedan was parked to the left in an empty row of spaces. Two men stood next to the back end. One wore a suit and had a leather satchel looped over his chest. He looked irritated. He must be Carter Hale, Skeeter's attorney.

I recognized the other man, dressed in jeans and a navy T-shirt, instantly. Then again, it was hard not to with Jed's tall, muscular frame. He had an imposing presence, and he'd scared me when I'd first met him. But I'd quickly gotten to know him and realized that while he could be a danger to his enemies, he took loyalty to a whole other level with his friends. My stomach flipflopped with relief and a warm feeling I couldn't name.

While in Henryetta, I'd let Jed make all the decisions about my safety, but I'd taken over for him after leaving. Sometimes I'd screwed up badly—like blindly trusting Wyatt in the beginning—but I'd learned to trust my gut. Seeing Jed there, though, watching our van with wariness, part of me wanted to cede that control to him again.

Wyatt released a snort. "Let me guess, that big guy is Jed."

"Yep," I said, trying to rein in my excitement.

Wyatt scowled, his hands tightening on the steering wheel.

"Trust me, Wyatt," I said quietly. "This trust thing is a two-way street."

"Only you haven't proven you trust me, have you? It's been all me."

I placed my hand on the back of the passenger seat. "Not true. I left Marco that non-incriminating voicemail message, and I didn't tell Jed I'd been kidnapped. But yeah, I know doing this is a lot bigger, so that's what makes it even more of a test." I put my hand on his shoulder. "I don't plan on escaping with Jed, if for no other reason than that he'd insist on helping me, and I refuse to put him in danger."

He slowed down, coming to a stop in the middle of the parking lot row.

Shit. He was going to turn around. I had to stop him. "Wyatt," I said, my voice firm, "I can assure you that if you turn around and leave, Jed will hunt us down, and you won't like what happens when he catches you."

Grumbling under his breath, he started driving again, then parked next to the sedan, leaving an empty space between the cars. I slid the back door open before Wyatt had even pulled to a complete stop, hopping out and restraining myself from running to Jed.

But he showed no such restraint, closing the distance between us and engulfing me in a tight hug.

I clung to him, blinking back tears. I'd missed him and my other friends in Henryetta. I'd missed the security they'd offered and the protected isolation of Rose's farm. Jed had done so much to help me, for absolutely no reason at all, and by setting up this meeting he was *still* helping me.

His arms still around me, he leaned back and studied my face. "How are you?" He lifted a hand and flipped my hair. "Still brunette and short, huh? Neely Kate thought you might go red."

"I doubt I could pull off being a redhead." I touched my hair self-consciously and shrugged. "I don't love it, but it seemed like the best way to not be Caroline Blakely."

His gaze shifted to the van, and his guard went up as Wyatt walked around the back.

"Who's your friend?" Jed asked with an edge.

I wasn't sure why he seemed so guarded, but then again, Jed had always been perceptive. He could probably tell I wasn't so sure about Wyatt.

I stepped to the side and gestured between them. "Jed, this is Wyatt Drummond. Wyatt, this is my good friend Jed Carlisle."

Wyatt and Jed stared at each other for several long seconds, neither offering the other their hand.

"*Well,*" Carter announced sarcastically, straightening his tie. "While this reunion and introduction have been extra fun, we need to get moving."

I looked up at Jed, suddenly nervous. It was easy to be brave when discussing the idea of visiting a hardened criminal in prison so I could ask him to give me confidential information he wouldn't tell anyone else, but now I'd have to actually follow through.

Jed smiled down at me, cupping my upper arm. "I can't go in with you. It's just you and Carter, but if Skeeter gives you trouble, just tell him Jed said 'hummingbird,' and he'll cooperate."

"Hummingbird?" I asked in confusion.

"He'll know what it means. Also—" he stepped away and opened the back door. "Neely Kate wasn't sure if you'd have the appropriate clothes to pose as his assistant, so she sent a bunch of clothes and shoes to help." He looked me up and down. "I think the dress will work, just add a jacket and swap out the sandals for heels." He reached inside and pulled out a black women's suit jacket and a navy blue one. "I don't have any fashion sense—" A sheepish grin spread across his face "—much to Neely Kate's dismay, so you'll have to figure out what's best."

I took the navy one, figuring it would go better with my light blue dress. I shrugged into it as he pulled out two pairs of pumps, one nude and the other black. Leaning against the back of the sedan, I slipped off my left sandal and reached for the nude pair.

Jed handed me the left one and then the right. After I slipped them both on, I straightened and tugged on the lapels of my jacket. "Do I look good enough?"

"Yeah," Jed said. "You look great."

"You look ten times more professional than my actual receptionist, so consider that a win," Carter said snidely. "Now let's go. I'd rather get my ass-chewing over with."

Which meant Skeeter wouldn't be receptive to our visit. I'd have my work cut out for me if I hoped to get anything out of him. I always had Jed's "hummingbird" as backup, but I hoped to do this on my own.

I nodded. "Okay."

"Wait," Jed said, reaching into the car again before standing with a small card in his hand. "You'll need this."

He reached out and I took the card, then stared down at my old driver's license. It felt unreal.

"There's no active search for Caroline Blakely," Jed said. "And Carter paid to have a fast background check done, so you'll be able to get in okay."

"How'd he manage that?" Wyatt asked in a flat tone as he leaned his shoulder into the back of the van.

"Money talks," Carter barked, then gestured toward me. "Let's *go*."

I glanced down at my license again, taking in the face of the woman I'd been a year ago. Blonde. Long-haired. Sweet. Innocent.

Oblivious.

I hoped I was as different from her as I thought I was.

Carter snapped his fingers at me, and I started walking, too nervous to tell Jed or Wyatt goodbye. We fell into step together as we headed for the entrance.

"Jed was worried you wouldn't look enough like your photo," Carter said as we walked at a brisk pace. "He suggested you wear a wig to match your license, but I assured him it won't be a problem."

"Does Skeeter know we're coming?" I asked, my stomach flipflopping.

"He's been told I'm coming to see him regarding his case, and I *do* have information, so that's good news, but he doesn't know about you joining me. He'll likely be pissed we're 'wasting' his time"—he used air quotes—"although it's not like he has a busy schedule,"

Carter said with a short laugh. "But for some reason, he doesn't like unplanned visits." He shrugged. "Then again, he doesn't seem to like visits all that much in general."

"Do many people visit him?" I asked as we approached the entrance.

"No, but that's because he refuses to see anyone. Jed included. I'm just about the only person he's seen since he was arrested a month ago."

"So he's not going to be happy to see me," I said more as a statement than a question.

He laughed again. "Understatement of the year."

CHAPTER TWENTY-THREE

CARLY

Twenty minutes later, after we signed in and were searched for contraband, we were led through multiple locked hallways and then ushered into a small room. It had several small windows near the ceiling, about six feet off the floor, but they were only about a foot tall each. Nothing big enough for a grown man to crawl through. The walls were a dull gray and a table with several chairs was placed in the middle. The table was bolted to the floor, and a round metal bar ran the length of one side, presumably so prisoners could be handcuffed to it.

"Hang tight," said the guard who'd escorted us. "They'll be bringing him in shortly." Then he walked out and locked us in, the sound of metal against metal stirring a deep, primal fear within me.

"Thanks," Carter said absently, even though the guard couldn't hear him, as though it was out of habit and not actual gratitude. He sat on the side opposite the metal bar and set his legal pad on the table. He picked up the pen and started writing. "Have a seat, Caroline."

It felt weird to hear someone unironically call me Caroline, but I sat in the chair next to him without comment. I was starting to have major second thoughts.

The sound of his pen scratching on paper filled the concrete-encased room, louder than it should be. I was forcefully reminded of Edgar Allen Poe's *The Tell-Tale Heart.*

Was I losing my mind?

After the decisions I'd made over the last twelve hours, not to mention the ones I was about to make, there was a strong argument to the affirmative.

"I suggest you get your nerves under control," Carter said, still writing, just like the heart in the story had kept right on beating. The heart was the narrator's conscience. Maybe the scratching pen was mine. "Skeeter Malcolm doesn't suffer fools lightly. And if you act like a scared mouse, he'll eat you in one gulp." He looked up with a cryptic grin, and the pen stopped. "Figuratively speaking, *of course.*"

I pulled back my shoulders and drew in a breath. "Noted."

He sat up straighter and set down the pen, turning at the waist to face me. "One more piece of advice: Try not to mention Rose Gardner. That's a bit of a sore subject with him." His face contorted into a pain-filled grimace. "If you'd planned on playing the sympathy card by saying you're Rose's friend, don't. You're better off using the Jed angle."

"What does hummingbird mean?" I asked.

"I have no idea," he said, turning back to his pad. "Those two have more history than a family bible. But I doubt we'll find out so don't press your luck."

The door opposite us opened and a guard led in a man in an orange jumpsuit. I'd never met Skeeter Malcolm in person, but I knew it was him. The real biological father of Rose's baby. With dark hair and a short beard, he was just as big as Jed. His dark gaze landed on me as the guard led him to the table and locked his cuffs to the bar.

"That's not necessary," Carter said, looking up from his writing.

The guard gestured toward me. "Gotta do it with her in here."

Carter started to say something, but Skeeter interrupted. "It's fine. This will be a short visit."

The guard shrugged, then stepped out of the room, shutting the door behind him.

"What's she doin' here?" Skeeter asked Carter, his voice tight.

"We'll get to her in a moment," Carter said, placing his palm on the notepad and staring into his client's face. "I have news on your case."

Skeeter shook his head in disgust and sat back in his chair. "I don't give a fuck about my case."

"Well, fortunately for you, I *do*," Carter said in a cheery tone. "Since the FBI reneged on their deal over a technicality, the US district prosecutor seems open to dropping the charges."

Skeeter waved him off and turned his attention to me. "Caroline Blakely. I have to say I respect your ingenuity in arranging a meeting."

So he knew who I was. Then again, that was no surprise from a man like him. He'd likely had me investigated while I was living with Rose last year. No one could accuse Skeeter Malcolm of being stupid.

I hid my fear and gave him a dark stare. "Rumor has it you're a hard man to meet, so I had to improvise." I could have told him that Jed had arranged all of this, but I'd prefer to let him think I was wily.

Carter snorted but kept quiet.

"Maybe there's a reason for that," he grunted. "But I have to admit I'm intrigued, so I'm not going to kick you both out. What do you want?"

"Who says I want something?" I said, keeping perfectly still. "Maybe I'm here to offer *you* something."

Amusement filled his eyes. "You didn't come all this way and arrange this visit to offer me something out of the kindness of your heart." His eyes turned cold. "So let's cut the shit."

I held up my hands in surrender. "You're right. I do want something, but I wanted to let you know that I'm willing to offer something in return."

He sat back, his arms slightly outstretched in front of him due to being tethered to the table. "I'm listening."

"I confess I don't have a lot of information about my father's illegal business dealings, but this is a pit stop on my way to Dallas to confront him. But if you need something from him, I'm willing to get it for you before I put him away."

He studied me with a blank face and then burst into laughter.

I continued to watch him while Carter stopped writing and sat back, studying us both.

"You think you can just waltz into your father's life and take him down?" Skeeter asked, still looking amused, but I had a feeling his good humor wouldn't last long.

"Look," I said, extending my upturned palm onto the table, "I know you don't think I can do it, but I *am* going to get proof he committed at least one murder, and I *will* have him put away."

He pressed his lips together. "It would be a hell of a lot easier to just kill him."

"I'd rather not resort to violence if I can avoid it."

He shook his head, obviously disappointed in my response. Or maybe I'd just shown him I was as naïve as he'd expected me to be. "The only things someone like your father responds to are lawsuits and violence. You planning to slap him with a civil suit for threatening to kill you?"

I wasn't surprised he knew that my father wanted me dead. Several people I'd grown close to with ties to Skeeter had known. "We both know I can't do that, and even if I could, it wouldn't hurt him. Not the way I want him hurt."

His gaze pierced mine. "Do you have any teeth to your bite, Caroline Blakely, or are you the toothless puppy who showed up in Henryetta a year ago?"

"I'll freely admit that one year ago, Caroline Blakely was a scared, helpless woman running for her life. But I'm Carly Moore now, and I'm not helpless, and I'm not running." I set my forearm on the table and leaned closer to him. "Caroline was a Golden Retriever, but I'm a Doberman with a nasty bite. My father is going down."

He studied me, then said, "You have gumption, I'll give you that." His head tilted to the side. "What makes you think I know anything?"

"I know you've met with my father and that you know things about his business. And we both know the bust you set up provided the evidence to arrest him, but for whatever reason the feds didn't go for it."

His gaze took me in for a few seconds before he said, "Let's just say the bust didn't go down as planned on both ends. Your father set me up."

This was good. Really good. I could work with this. "So you have a vendetta to settle with him too."

A lazy grin tipped up the corners of his mouth. "Caroline, you may be a Doberman, but I'm a Great White Shark."

"But you're in prison and don't seem in any hurry to get out, and I'm going to see him *now*. If you want to be part of that, now's your chance."

"For a price," he said in a lazy tone. "Of course."

"You got a shitty deal with the FBI," I said, ignoring his statement. "You helped them make a bust they couldn't pull off on their own, then they screwed you because it didn't go down exactly as it should have. But you're an intelligent man, a schemer—and I mean that as the highest form of praise—which means you saved something. You have something else the FBI wants, you just haven't told them yet."

His brow wrinkled. "If I had information, why would I hold onto it? Why not use it to get the hell out of here?"

Why indeed? And then it hit me like a ton of bricks. "Because my father has something you want, and you can't get it if he's in prison."

"Let's entertain your little fantasy," Skeeter said, shifting in his seat. "Let's say I *do* have something that can get me out of here. Why the fuck would I tell *you*? Why not get out and get it myself?"

He had a point. He could go after my father himself, but he hadn't. Why? "I don't know," I admitted. "But there's no denying you want revenge against my father, and when I'm finished with him, you'll lose your chance. I'm being generous and offering you an opportunity."

Skeeter stared at me like I was delusional, then burst into laughter. "You really think you can bring your father down on your own? *You?* I saw you when you came to Henryetta last summer, Caroline. You were scared of your own fucking shadow. You've lost your mind if you think you can neutralize him."

"Maybe I *have* lost my mind, but I assure you," I spat out, "I *will* bring him down, and I don't plan on doing it alone."

"Let me guess," Skeeter said derisively. "You've got a boyfriend helping you."

"No," I said, my voice hard. "I have a vile man who won't think a thing about breaking whatever laws are necessary to get to him. Wyatt's after a reward, and he'll do anything he can to get it."

Skeeter's brow lifted. "Including screw you over?"

"Possibly, but he knows I'm not after my father's money, and he's had a thing for me in the past. I'm counting on his feelings for me to keep him from totally screwing me over."

His eyes darkened.

Shit, I suspected I'd just reminded him of his relationship with Rose.

"My boyfriend is a deputy sheriff," I said. "He wants to help me, but he has too much integrity to do what it will take to get to my father. But Wyatt will. I know he's not trustworthy, but he doesn't realize I'm not either."

That caught his attention. "You plan on double crossing him?"

"Not if I can help it, but I will if necessary."

He studied me for a moment, like I was a word jumble that needed sorting. "Why are you doin' this at all? Why not just spend the rest of your life as Carly Moore making babies with that deputy of yours?"

"Because I've spent the last year looking over my shoulder, wondering if and when my father would find me. Because I'm pretty damn sure he discovered my alias yesterday and now has a general idea of where I've been living. If I'd stayed in Drum, he would have found me in a matter of days." Frustrated, I put both hands on the table. I needed him to believe I could do this. "Yeah, I ran in the beginning, but I'm not a scared little girl anymore, Mr. Malcolm, and I refuse to spend the rest of my life running from him," I said, narrowing my gaze. "But *he* still sees me as a scared little girl. He'll never expect me to show up at his doorstep on the offensive. That alone will be worth seeing the shock in his eyes."

He leaned his face closer and nodded, lowering his voice. "And then you put a bullet in his head, Caroline. You sure as hell don't try to play by the goddamned rules of justice. Men like your father twist those rules to their own will. You *will* lose."

I held his gaze. "Maybe so," I said assuredly, "but death is too good for him. My mother died in a car accident when I was eight. I thought it was an accident at the time, but he had her murdered. He'd intended for me to be killed in that accident too. I wasn't with her, though, so he settled for making my life absolute hell. Until he could marry me off to a man he'd paid to spy on me. The guy who'd pretended to be my friend all while he was sabotaging all my relationships." Anger burned deep in my chest. I'd lived with this knowledge for a year, but now that I was so close to getting my revenge, it seemed to burn hotter. "Randall Blakely will come face to face with the girl he discarded. He'll know it's me avenging my mother's murder."

"This isn't a game, little girl," he sneered. "Leave this to the grownups."

"You're right. It's *not* a game," I forced out. "I know you see me as the scared woman our friends helped last year. But I'm not her. The night I arrived in Drum, Tennessee, a teenage boy was killed in front of me, and I promised his grandfather I'd avenge his death. And I *did* make his murderers pay, only it didn't feel like enough, because I suspected that Bart Drummond, the town patriarch, had sanctioned his killing. I vowed to prove it. It took me eight months, but that bastard's sitting behind bars with murder charges he won't be able to bribe his way out of," I said through gritted teeth. "I've dealt with your equivalent in Drum, a drug dealer the whole town fears, but he respects me. We've worked together to find his kidnapped girlfriend, suss out his competition, and take down a dirty sheriff's deputy." I lowered my head to hold his gaze. "I'm not scared to get my hands dirty and do what needs to be done to get justice. But, no, I don't plan to kill Randall Blakely, because that would be too good for him. I want him to spend the rest of his miserable life behind bars, knowing the daughter he treated like trash was the one to do it."

A slow grin cracked his lips. "By damn. You're a live wire."

I sat up straighter. "I can do this, but I don't want it to take eight months like it did with Bart Drummond. It has to be done within days. I'm not hiding anymore. I need to take him by surprise while he thinks he's closing in on me in Tennessee."

"You don't plan to do it alone. You plan to have help," Skeeter drawled. "Your non-boyfriend."

"Believe me, Wyatt Drummond has done plenty of his own underhanded bullshit. He's up to the challenge."

His brows shot up. "Drummond? Any relation to Bart? Why would he turn on his family?"

Shit. "Wyatt is Bart's son, but he hates him. He tried to make his own deal with the FBI to put his father away, but I got Bart arrested first."

He gave me the side-eye. "Wyatt must be pissed."

"He is, but he's turned his attention to my father. He thinks he can get a new deal out of it."

"And you're teaming up together?" He waved his hand, the chain rattling. "Why do I think there's a whole lot of context you're not telling me?"

"It has nothing to do with you." I drew in a breath, hoping I hadn't burned a bridge, but he didn't want to discuss his love life and neither did I. My back stiffened. "Don't underestimate me, Skeeter Malcolm."

He laughed again, but this time, it didn't sound like he was laughing *at* me. "Like I said, you've got gumption. But gumption alone won't bring the man down, and I'm not sure it's a safe bet to hand you something to help you with your witch hunt." A grim smile spread across his face. He pressed his lips together, giving me a sympathetic look. "Sorry, Carly Moore. Can't do it."

"Then whatever it is you want from him will be lost forever."

He looked irritated, like I was a pesky fly that kept buzzing around his head. "You may be a live wire, but the chances of you succeeding are slim to none. Your father is one of three men who ruled an international crime syndicate, and the only experience you have is from taking down a two-bit player in some backwoods Appalachian town."

I ignored the fact that he seemed to know I'd been living in the mountains. Maybe it was a lucky guess. "Bart Drummond was more than a two-bit player."

"Doesn't matter," he said with a wave of his hand, "never heard of him. I wish you luck, but you won't be getting any help from me."

"We both know you want something. On the off chance that I *do* succeed, what harm is there in you telling me what you want?" I asked. "Indulge me."

His jaw tightened. "I don't see the point of wasting more time."

"What else do you have to do right now?" I asked. "Get back to arts and crafts? Or are you worried about losing some of your gym time?" I waved my hand at him. "You look plenty muscular as it is. I think you can miss a day at the bench press."

His eyes widened as though silently asking if I had a death wish.

Maybe I did.

"Let's say I skip having my father arrested and go with your suggestion to kill him. You'll lose your chance to get what you want from him," I said. "Are you really willing to risk it?"

He studied me for several seconds, looking like he wanted to murder me with his eyes. "The location of the Rucker locker."

I shook my head slightly. "What does that mean?"

"You don't need to know that part. I just need the location."

I narrowed my gaze. "Don't you think a little context might help?"

A sly grin lit up his face. "What would be the fun in that?"

"I know you underestimate me, Mr. Malcolm, but I *will* see this through. I can promise you that."

He leaned forward. "Everything has a price, whether you realize it or not at the time." He sat back again, some of his menace fading. He looked more like the businessman Rose had claimed him to be at heart. "Besides, what good would any of it do me if I'm stuck in here?"

Which raised a good point. Why would a ruthless man like Skeeter Malcolm *want* to be in here? Why wouldn't he use the information I knew he was sitting on to get out? It just didn't make sense.

Until it did.

Just like Wyatt was working the system with a Plan B, maybe Skeeter Malcolm was doing the same thing. He was in here for a reason, one only he understood, but he saw his incarceration as temporary. That meant he really would want that location.

I actually had a shot at getting real information out of him, but I had to play it right.

I shrugged. "Somehow I don't think you'll spend the rest of your days in lockup, so an insurance plan is a good bet. I'll get you the location of the Rucker locker for you, but you have to give me more to go on. And you have to give me something I can use to get to my father."

A sneer spread across his face. "And there it is. Like *I* said, nothing's free. And if I'm getting out like you think I am, maybe I plan to go get the information myself."

"Then you'll be sorely disappointed that you missed your opportunity. Are you willing to risk it?"

Skeeter's face hardened. "Maybe I'll take my chances."

"Rose always considered you a smart man," I said flippantly. "Maybe she was blinded by—"

Skeeter tried to get to his feet in a rush, jerking on the table with a hard yank when his chain only let him partially rise. *"Don't you talk about her!"*

I knew he loved her and that she'd loved him at one time. Bringing it up hadn't been smart, but it had just come out, and I couldn't deny it had gotten a reaction out of him. Just maybe not the reaction I was looking for. Still, some reaction was better than none at this point. At least he wasn't fake-playing nice anymore.

Carter held out his hands between us. "Okay, now. Enough of that." He pointed a finger at me. "You were warned." Then he turned to Skeeter. "You need to control your temper."

Skeeter gave the chains a hard yank, staring at me like he wanted to use it around my neck, but we both knew it was for show. It didn't scare me. Mostly. At this point, I had nothing to lose.

"Look," I said. "I only need enough information to make him take me seriously. If he thinks I know things I shouldn't, I can get him to admit to things I don't. Surely the FBI will arrest him then."

"You mean take you seriously besides the gun you better have pointed at him when you walk through his door?" Skeeter spit out. "Won't that be enough?"

"I need him to know I've changed."

His gaze narrowed. "You don't need me for that. You seem to be holding your own with me just fine."

Maybe he was right. Maybe I'd just show up at my father's house and catch him sitting in his leather chair by the fire in his study. Then hold a gun on him and make him admit that he'd arranged for my mother's car to be tampered with, and that he'd done the same to her parents. And that he'd intended for me to die too. And then I'd tell him I knew he'd paid Jake to spy on me and orchestrate my life so that I'd be broken enough to marry him. All so Jake could bypass some arcane rules and become part of my father's criminal network. I just needed to capture it all on tape.

Maybe I didn't even need Wyatt. Maybe I just needed to walk in like I owned the place.

A wry grin lit up Skeeter's eyes. "You may think you can get away with putting him behind bars, but he'll still have influence. If you really want to make him pay, I'd rethink your policy on murder."

"That would make me just like him."

His brow lifted and he said thoughtfully, "Maybe you are. Maybe that's not necessarily a bad thing." He glanced over his shoulder and shouted, "Guard! I'm ready."

"Wait," Carter said, holding up a hand. "You and I aren't done yet."

"Sorry," Skeeter said with a wink. "Gotta get back to my arts and crafts. I was in the middle of making a popsicle stick log cabin."

The door opened and the guard walked in.

"Skeeter!" Carter protested.

Skeeter ignored him and held up his hands for the guard to unhook him from the table.

Carter released a snort of disgust and shoved his legal pad and pen to the middle of the table as the guard led Skeeter to the door.

Then Skeeter stopped at the doorway and turned back. "If you really want to get under Blakely's skin, tell him you know about the Carpathian deal—that was the deal we were working on at the end—and say he was the one to fuck it up." A grin spread across his face. "Rumor has it that'll set him off. He's bound to spill some information you might find useful."

I held his gaze. "Thank you."

"If you want to thank me, then get me that location." He waved a hand as though trying to salute. "Good luck, Caroline Blakely. You're gonna need it."

CHAPTER TWENTY-FOUR

CARLY

When we exited the building, Jed and Wyatt were still standing next to their respective cars in a sort of stand-off. Jed saw us and stepped away from the car, advancing toward us.

I felt weirdly numb, as though talking to Skeeter had unlocked something feral and ugly inside of me.

Maybe it was because he'd confirmed my worst fear—the only way to truly make my father pay was to kill him. And if that really was the case, then I definitely needed to keep Marco out of it. Even if I got away with it, I couldn't drag him down with me. He'd never be able to live with himself.

"How'd it go?" Jed asked when I was a few feet away. "Did Skeeter give you anything?"

"He did at the end," Carter said, walking past us. "She monopolized the entire conversation, and I never got a chance to discuss the latest developments in his case."

"You can come back and talk to him next week," Jed said, keeping his gaze on me. "How pissed was he when you used hummingbird?"

"She didn't have to use it," Carter said as he opened the back door. He tossed his briefcase onto the seat. "He told her all on his own."

"He did *what?*" Jed asked in shock.

Carter slammed the back door, then turned to face us. "Yep. She conned him into believing she could get the evidence to put her father away and he relented."

"She *can* put him away," Wyatt said, his back still resting against the van, his arms crossed over his chest. "If anyone can, it's Carly."

Jed gave him a dark glare, but Carter turned to him with interest. "What's in this for you, Lover Boy?"

"What the fuck business is it of yours?" Wyatt snapped.

"Because my client has given Ms. Blakely a task, and I need to know if you're going to interfere with that."

Jed's shoulders jerked. "A task? What kind of task?"

"That's client-attorney privileged information," Carter said.

"Carly's not an attorney," Jed pointed out in an ominous tone.

"No, but he told her in my presence. Seeing how reluctant he was to give it to her; I'm guessing she needs to guard the information with her life. Literally, because her life just might depend on it."

The warning stare Carter gave me suggested my father wasn't the only one I needed to fear if Skeeter's information was shared with the wrong person.

"Carter's right," I said, lifting my chin. "It's confidential information I'll use when it's time, and not before." I sure as hell wasn't giving it to Wyatt. He'd hear about it when I mentioned it to my father.

If he was still with me by then.

"You think I'm going to accept that bullshit answer?" Wyatt asked, pushing away from the van.

"You will if you know what's good for you," Jed said, taking a step toward him.

"I put a hell of a lot on the line to bring her here," Wyatt said, pointing to the asphalt at his feet. "She's gonna tell me what she found out."

Jed pulled his wallet out of his back pocket and pulled out some bills. "Then let me pay you for your time and trouble, and we'll take it from here." He held out what appeared to be several hundred-dollar bills, but Wyatt shoved his hand away.

"I don't want your damn money," Wyatt spat, turning his attention to me. "We had a deal, Carly."

"We still have a deal," I said. "So calm the fuck down."

"You're not thinking about actually leaving with this lowlife, are you?" Jed demanded in disbelief.

"It's okay, Jed. I'll be fine."

"You don't think I know that you're in trouble?" Jed demanded. "I heard it in your voice when I called you. Neely Kate and Rose heard it too."

"I'm *fine*," I insisted.

"*We* can help you with this, Carly," Jed said. "You know we can. And you can *trust* us. You can't tell me that you trust *this* guy." He gestured toward Wyatt.

I put my hand on Jed's arm, staring up at him. "Wyatt's right. We made a deal."

"Where's your boyfriend?" Jed asked. "Marco. Rose told us about him. Where's he in all of this? Does he know you're with this asshole?"

My heart lurched at the mention of Marco's name. "I can't involve him in this. What I'm planning on doing isn't on the up and up, and Marco's a— "

"Deputy sheriff," Jed finished for me. "But while you've been inside, I've been trying to chat with your friend here. When he refused to talk, I looked him up." His brow lifted. "Wyatt Drummond from Drum, Tennessee, no?"

My mouth dropped open. I hadn't told him Wyatt's full name.

"Don't look so surprised," Jed said. "All it took was a bit of Google searching along with what you'd told Rose. Plus, I have access to deeper dive searches. I saw his DUI charges and incarceration."

"I served my time," Wyatt said. "I don't hide it."

"No, but there're a lot of gaps in what I *did* find, most of them questionable."

"I don't give a rat's ass what you think about me," Wyatt said, taking a step toward me. "You saw Malcolm, Carly, and now it's time to go. That was our agreement."

Jed stepped in front of him, puffing up his chest. "I don't think so."

"Jed," I said carefully. "It's okay."

"The fuck it is."

Carter held up his hands. "I know you don't want to hear this, Jed, but Skeeter gave her the information partly because she told him she was working with Drummond on this. She made a compelling case."

Wyatt's glare intensified while Jed's scowl deepened.

"Like I said, he was reluctant to give her the information," Carter went on. "In fact, I suspect if he thought you were any part of it, he would have kept it to himself."

Jed's jaw clenched as he turned to Carter. "I'm not letting her go with him. The phone call she made to set up this meeting was sketchy as hell." He pointed to Wyatt. "He's untrustworthy."

"Jed," I said firmly, "while I appreciate your concern, this is my decision, not yours. I'm leaving with Wyatt."

"Where's your wallet, Carly?" Jed asked. "Where's your purse? Why don't you have any ID?"

I hesitated. "I had to leave town quickly."

"Why?" Jed asked, his brow lifted.

"That's none of your fucking business," Wyatt said, then reached out for me. "Carly, let's go."

"I said I was leaving with you, Wyatt, so calm the fuck down," I snapped. "You're not helping."

Jed gently grabbed my upper arm. "I know you're in trouble. Let me help you."

I released a bitter laugh. "I've been in trouble most of my life, but if you're asking if I'm in over my head with Wyatt, then the answer is no. We're working on this together. He thought it was too risky for me to see Skeeter, but he didn't stand in my way."

"Why is it risky?" Jed asked, his voice even with the confidence of a man who had an ace in the hole. "Is it because Wyatt Drummond's currently wanted for kidnapping a woman from Drum, Tennessee last night?"

Wyatt stomped toward me, but Jed held up his hand. Thank God, neither one of them pulled out a gun. Then again, the guards stationed in the guard towers probably would have stopped them. In fact, I wouldn't be surprised if this entire encounter were being observed.

"Trust me," Jed growled at Wyatt. "You *will* lose if you try to go through me." He turned to me. "Carly, explain to me why you're leaving with a man who kidnapped you at gunpoint, tied you up in a van, then shot at your boyfriend."

Shit. Why was I surprised he'd discovered all of that from his sleuthing? The real question was why he hadn't addressed it the moment I'd walked out of the prison.

"Seems too soon for Stockholm Syndrome," Carter drawled.

I grimaced. "I know it looks bad."

"You *think*?" Jed said.

"Wyatt and I have worked out a truce." I squared my shoulders. "I know what I'm doing, Jed."

"*Do* you?"

"I've been on my own since I drove away from you and Neely Kate and Rose last October. Sure, I've made new friends in Drum, and I've found Marco, but I've also learned that the only person I can depend on is myself and my intuition. I can handle this."

"This is asinine, Carly," Jed protested.

"I know," I agreed. "And yet, it's still my decision."

He sucked in a breath of frustration, then blew it out. "Fine. I'm comin' with you."

"No fucking way," Wyatt grunted.

"Jed, I can't let you do that," I protested. "I could go to jail for this."

"What the hell are you plannin'?" he asked in surprise.

"Whatever I need to do to get him to admit to murdering my mother."

He ran a hand over his head, and turned away, taking a few seconds to get himself together.

I placed a hand on his shoulder. "Jed."

He slowly turned back to face me.

"I love you and Neely Kate too much to let you come. I'll call you as soon as it's done so you know I'm safe, but you can't be part of it." I grabbed his hand and squeezed it. "Trust me."

He searched my face. "I trust you. It's *him* I don't trust."

To my surprise, Wyatt kept his mouth shut.

Finally, Jed cursed under his breath, then said, "I sure as hell hope you know what you're doin'."

My mouth lifted into a tired smile. "I hope so too."

"Not exactly confidence-instilling."

"And yet brutally honest."

He wrapped me up in a hug, squeezing tight. "You call me the moment you realize this asshole is screwing you over."

I chuckled against his chest. "Okay."

"I'm giving you a phone," he whispered in my ear. "You use it to contact only me. Then it will stay clean, and I can find you if necessary."

I pulled back slightly and looked up at him. "You planned for this?"

"Skeeter didn't get to where he was on his own. I was paid to plan for all contingencies."

I hugged him again. "Thank you, Jed. I owe you more than I can ever repay."

"No more talk about that. Just come see Neely Kate and Daisy when this is all done. And Rose and Hope too."

"I will."

He released me, then headed to the back of the car. Pulling out a small women's purse, he unzipped it and took out an iPhone. "This is *your* phone." He shot a glare at Wyatt. "I expect her to answer when I call. If she doesn't, I'll be coming for you."

"It's not up for discussion, Wyatt," I said, not bothering to look at him.

Jed handed me the purse, and I looped the strap over my shoulder, surprised by how heavy it was. "There's a couple of thousand in cash in there and a prepaid debit card with five thousand dollars loaded onto it. You can use it to stay in a motel. Keep using your Caroline Blakely identity. I know it's dangerous, especially the closer you get to Dallas, but there's a BOLO out for Carly Moore and Drummond there. They know you're headed to Texas, so *you* might want to drive in case you get pulled over." He glanced at my hair. "And you might consider bleaching your hair back. Even if you go by Caroline, you look enough like the photo they're circulating of Carly Moore that they might detain you."

I grabbed the phone and took the bag. "The van's stolen, so my hair will be the least of our concerns if we get pulled over."

"Then don't get pulled over, and when you get to Little Rock, get a rental car. I'll reserve one for you and text you the details." Jed turned to face Wyatt. "I expect her to give me multiple updates. At least three times a day, so don't try any bullshit to keep that from happening."

Wyatt's response was a glare.

Jed continued, "My number's preprogrammed in there as well as Neely Kate's and Rose's."

I clutched the phone to my chest. I wouldn't call them, but it was nice knowing I could.

"Jed," I said hesitantly. "There's a chance that Marco will call you or Neely Kate or Rose. He'll want to know if you've heard from me. If he does, tell him I saw Skeeter and that I'm okay." My voice broke. "Tell him I love him, and I'll call him when this is done."

Worry filled his eyes. "Carly..."

I shook my head, determined to not break down. "It's better this way. I've already asked Marco to cross too many lines. He has to stay out of this."

Jed paused, then nodded. "I'll tell him."

I gave him one last hug, and he whispered in my ear, "There's a loaded gun in the purse in case you need it. Extra bullets too. The gun's untraceable, so don't be afraid to use it." Then he dropped his hold on me and took a step back.

I stared at him in shock.

"I'll *always* have your back, Carly," he said softly.

Tears filled my eyes. He had no reason to do this, and yet here he was, giving me a way to contact him, and the means to protect myself and obtain food and shelter.

"I'll pay you back. I swear."

"The best way to pay me back is to come see us as soon as this is done."

"I will," I said, my voice breaking. "I swear."

"Then you better get goin'. Dallas is a good six hours from here."

I wanted to give him one last hug, but I stopped myself, worried I'd change my mind. Instead, I walked over to Wyatt, holding out my hand, palm up. "Keys," I said firmly. "I'm driving."

But Wyatt looked like he was waging his own internal war. I was sure he didn't like Jed schooling him like he was a naughty little boy, let alone me holding back information and taking charge. The fury in his eyes suggested he would rather strangle me, but he just said, "Already in the ignition."

"You're riding in the back this time."

He muttered under his breath but got in as I walked around to the driver's door. I took a breath to settle my nerves.

Think about this long and hard, Carly. It's not too late to turn back.

But there was no turning back. I'd started down this road the moment I'd decided to find justice for Seth. Because my mother deserved justice too, and so did I, and I'd be damned if I let that man take it from us.

CHAPTER TWENTY-FIVE

CARLY

I drove out of the parking lot and headed back toward the highway, trying to put on a disaffected front for Wyatt. He was sulking, but I sure as hell didn't plan to fill him in on what Skeeter had told me, let alone tell him about the gun in the purse I'd tucked under the front seat. I still trusted that he wouldn't hurt me, but if he tried to tie me up again, I wouldn't hesitate to use the gun to defend myself. Maybe I'd actually be able to use it this time, and shoot him in some non-vital location.

In the meantime, he'd learn the silent treatment didn't phase me. My father had frozen me out for most of my life.

I felt different after walking out of the prison, and it scared me. While I felt more determined, I also felt…cold. Like something in that gray room had hardened me and made me stronger, strong enough to do the job that needed to be done.

That was good, because I also needed to find it in myself to turn my back on Marco. He might forgive me if I did this, but he'd never forget. Maybe he'd never fully trust me again. I was scared to death that it would mean the end of us. That was the last thing I wanted, yet I couldn't ask him to stoop this low into the gutter. I loved him too much for that.

Wyatt...he was comfortable there.

I spent the next five minutes driving past fields full of soybeans and corn, heading toward the interstate. I opened the map app on my new phone and plugged in the general destination of Little Rock.

"You're seriously not going to tell me what happened in there?" Wyatt finally demanded, crossing his arms over his chest.

"Jesus Christ, Wyatt," I muttered, pressing the button to start the navigation on my new phone before I turned on my signal to get on the I-40 entrance ramp heading east. This would be a lot more convenient than using a paper map like we had to get to the prison. "How did I forget how pouty you are when you don't get your way?"

He jerked upright in his seat. "*Excuse me?*"

"You heard me," I said, merging onto the highway. "Every damn time you don't get your way, you pout."

"The fuck I do," he snarled.

I held up a finger. "When you found out I was talking to Bingham and wouldn't tell you why, you pouted." I held up another finger. "When I started hanging out with Marco, before he and I even thought about getting together, you pouted." I held up a third finger. "When you—"

"You're a fine one to talk," he scoffed. "You pouted plenty when I refused to talk about my father with you."

"Big difference," I countered bitterly. "You *promised* to work with me, and you *lied.* I gave you plenty of chances to come through, and when you didn't, I ended things. I never promised you shit with any of that stuff, and yet you pouted like a big man baby when you didn't get what you wanted." He started to protest, and I held up my hand. "Stop pouting."

He shut up and let out a big huff.

"I could so easily have ditched you back there," I continued, "and I didn't, so take that into consideration when you start getting pissed that things aren't going the way you want."

He snorted. "Like I would have let Jed take you."

I burst out laughing.

"What's so funny?"

I found his gaze in the mirror. "You really think *you* could have taken Jed? You're more delusional than I thought."

"Why the hell couldn't I take him?"

"Please," I said, rolling my eyes. "He helped Skeeter run a large criminal organization and was the man's personal bodyguard. He would have made mincemeat out of you." He started to protest, and I held up my hand again. "Stop. I realize your fragile male ego is bruised right now, but I have bigger issues to deal with than making you feel better about yourself."

"I could shoot you right now, Carly," Wyatt said. "Don't forget I kidnapped you at gunpoint and tied you up. I can do it again."

"And that was my mistake for believing you'd hurt me."

"You don't think I will?"

"This morning you said you wouldn't have shot me, so which is it?" I asked, getting pissed. "Either way you lied. Which one are you going to cop to?"

"Fuck you, Carly," he spat out.

"You'd like to, but that's not happening. Ever. This is purely business. As soon as we take care of my father, we'll go our separate ways and hopefully never see each other again."

He shook his head. "What happened in there? You came back a raging bitch."

"*Good.*" Lord knew I was going to need to be to get through what I had to do. If that meant I had to be Bitch Carly, like Wyatt had called me, then so be it.

"If you're so disgusted by me, then why didn't you go with Hero Jed?" he sneered.

"For all the reasons I told him. Do I think he's the better man for the job? Yeah, I do. But he's got a wife and a baby, and I'm not jeopardizing that for him. You seem intent on getting my father, and unlike you and *your* father, I'm willing to share my mission. You'll come in handy."

"I'll come in handy," he muttered in disgust. "You've got to be shittin' me."

"Plus, there's the information you claim to have."

"Claim?"

I was starting to second guess bringing him along at all. He was a risk, wanted by the police. I was just the victim of a crime. I could easily get Jed's rental car and go the rest of the way on my own.

Maybe that was the way it was supposed to be. I'd fled my father on my own. Maybe I needed to face him on my own too.

Only I couldn't exactly approach him by myself, with no backup. And I was far from sure he'd admit anything to me, even with the information Skeeter had given me. Even if I got the information, I wouldn't be able to get out without help. My father wouldn't willingly let me walk away. I'd need some firepower to help me leave. I wasn't sure even Wyatt was enough.

Talking to Jake first still seemed like a good option. If I caught him by surprise, he might tell me what he knew, especially if he didn't think I was a threat. Jake was exactly the sort of person who'd spill his guts to save himself, and if Wyatt's information was accurate, he was in trouble.

If I kept Wyatt around, he could be my protection.

Marco could do that.

He could, and he would, but I felt a bone-deep need to protect him. Just like he'd protected me from Carson Purdy and so many other things.

"How'd you get that information on Jake?" I asked, shooting a glance over my shoulder.

"Which part? The part about the inside trading came from the FBI, just like I told you."

"And the rest?"

He grinned. "I found that part out on my own."

"When did you start looking into him?"

He gave me a smug look in the mirror. "If you're not giving me information, then I'm not giving *you* information. How do you like that?"

"Like it's any different than how you usually handle things," I scoffed.

He shook his head, anger flashing in his eyes. I wasn't sure it was smart to provoke him, but I was pissed that he'd gotten me into this situation, separated from Marco and thinking it was better this way.

Still, I couldn't deny that it was in my power to change that, especially now. So I couldn't really blame Wyatt. Maybe all the things that had happened to me, from the backstabbing men to the threats to my life, had engulfed me in a slow boil, and the thermometer had finally hit the bitch stage.

After about a half-hour, Wyatt looked like he'd calmed down, so I said, "We need to come up with more of a plan."

"I already have a plan, Carly," he said in disgust.

"Well, you're gonna have to tell me what it is," I said, getting pissed. "I'm not going into this blind." He started to speak, and I held up a finger. "And don't you dare say you're not going to tell me."

"We confront Jake, then confront your father."

"Don't we need something more refined?"

"It involves some surveillance."

"So you don't have a plan," I said dryly.

"Let's just get to Dallas first," he said, sounding less angry. "We're both exhausted. We need to get some sleep before we tackle this."

"Where are we going to confront Jake? At work? At his house?"

"His house. Doesn't he work for your father? Going to his office doesn't seem like a good idea."

"We'll need to think about this because if we don't time it right, Jake'll warn my father that I'm in town." I glanced at Wyatt in the mirror. "We're looking for hard evidence, right? I think we should drop by Jake's house while he's at work tomorrow. We can talk to him *after* we search it."

"You want to break into his house?" he asked in surprise.

"I broke into *your* house and found evidence under the floorboards under your bed. Besides, I have the code to get into his house, and I doubt he changed it, so it's not really breaking and entering."

"Try telling that to a judge," he said sarcastically.

If Marco were with me, he'd insist on helping me with that too, which only proved I'd made the right decision leaving him out of this. "Got a better idea?"

He scowled. "No, but there's a slight problem. Tomorrow's Saturday. While your father might be a workaholic, something tells me that Jakey-boy prefers a cushier lifestyle."

I couldn't argue with that. "He plays golf. He joined my father's country club. In fact, they might be playing together tomorrow." A new idea was forming. "We'll need to start stalking Jake's house early tomorrow morning."

"How early is early?" he asked dryly.

"When was the last time you saw a sunrise?"

CHAPTER TWENTY-SIX

MARCO

Hank and I made it into Dallas a little before six. I drove the entire way from Memphis, with only a couple of stops at gas stations. Both times, Hank got out and went to the bathroom, then walked around outside, every step looking labored. On the second stop, I asked him if we needed to check his blood sugar.

"I've got it under control."

Only I wasn't so sure he did, so while he was purchasing some peanut butter crackers and a banana, I searched the small cooler in the back and only found one small vial of insulin and five syringes.

Was that enough? I hadn't seen him check his blood sugar, and he wasn't acting like himself. He'd been through a lot in the past few days—hell, the past year. Maybe I shouldn't have asked him to come. Maybe he'd be better off back in Drum.

He slept for the rest of the drive, even when we were in the thick of Dallas's Friday night rush hour traffic. I pulled into a hotel off the interstate and turned off the engine, and he kept right on sleeping. Suddenly worried that he'd gone into a diabetic coma, I shook his arm. He roused, grunting, "What?"

A breath of relief rushed out of me, but I sure as hell wasn't about to tell him I thought he was sick. "We're here."

"Where?" He struggled to sit upright and glanced around the parking lot. "Where the hell are we?"

"In Dallas. I figured we'd check into a hotel, then look up Duncan Morris. See if we can get an address for him so we can drop by for a chat." Although I had to admit that Hank looked like he needed to go to bed and sleep until morning.

"Yeah," he said, reaching for the door handle. "Sounds like a plan."

I grabbed the bags and headed into the lobby, both of us moving slowly. My leg was throbbing, and I was walking with a pronounced limp, while Hank looked like he was struggling to find the energy to use his crutches.

"Picked a fancy place?" Hank grunted, checking out the marble floor in the lobby.

"I wouldn't exactly call a Hilton a fancy place," I said in a low voice. "I picked somewhere that has room service." We needed a meal that wasn't greasy hamburgers and fries, one we wouldn't have to sit in a restaurant to get.

Since I hadn't made a reservation, we were lucky the hotel had a room with two double beds. I reserved it for two nights, hoping we'd find Carly and be ready to get out of here in two days. I couldn't entertain the idea of *not* finding her, but if I was honest with myself, I didn't know *how* to find her. Stake out Randall Blakely's house? For all we knew, Wyatt had already beat us to Dallas and turned her over. Maybe we should just show up on Blakely's doorstep and demand to see Carly.

"What's whizzin' through your head?" Hank barked as we headed for the elevator.

"I'm wondering if we should immediately go over to Blakely's and see if she's there. We can look into Morris later."

"Let's dump our stuff off first, then go park out front and stake out the place."

While I was glad he'd agreed, it also worried me. It meant he thought she might already be there too.

When we got out on the fifth floor, I considered just turning around and programming Blakely's address into my phone, but our room was only three doors down. We might as well drop off our stuff and go to the bathroom.

The padded carpet seemed to make it harder for Hank to shuffle on his crutches, proving to me that he was losing his energy fast. He needed a decent meal. Maybe I should order room service for him, then leave him at the hotel while I staked out Blakely's.

Two things stopped me. One, he'd kill me—perhaps literally—and two, I was pretty sure I was going to need his experience on the other side of the law to guide me. It wasn't like I could show up at Blakely's doorstep and flash my Hensen County Deputy Sheriff badge. Not that I was a Hensen County Deputy Sheriff anymore. I'd emailed them a letter of resignation from my phone before we'd left Memphis.

Bottom line, I needed Hank, but he was in no shape to play commando.

We'd just reached the door to our room when my cell phone rang.

I gasped. "Carly."

Handing the room key to Hank, I scrambled to get my phone out of my pocket, hoping and praying it was her. It was a 501 area code. Arkansas. Had Carly somehow gotten away from Wyatt and found her friends in Arkansas?

I looked up into Hank's face. "501 area code. She might have found a phone."

His face dark, he gave a stiff nod. "One way to find out."

My stomach twisted as I accepted the call and put it on speaker. "Marco Roland."

A man's voice responded. "Marco, you don't know me, but my name's Jed Carlisle. I'm a friend of Carly's."

I placed a hand on the wall to keep my balance. As far as I knew, Carly hadn't spoken to this man since leaving Arkansas. It couldn't be a coincidence that he was calling now. "Please tell me Carly's with you."

"I wish I could," Jed said solemnly, "but I couldn't talk her into letting me go with her."

I blinked hard. "*What?* You saw her?"

Hank held the key up to the door, opened it, then pushed me into the room.

I flicked on the lights.

"Back up," Hank said as we walked past the marble-encased bathroom. "Where did you see her?"

"Who's that with you?" Jed asked, turning wary.

I dropped the bags onto the first plump, duvet-covered bed. "Hank Chalmers. He took Carly in when she came to Drum."

"Carly told Rose about that," Jed said. "I heard you've been good to her. Thank you for lookin' out for her."

"We can do the kumbaya bullshit later," Hank grunted. "Tell us what you know."

"Carly called Rose and Neely Kate this morning at the nursery and asked them to set up a meeting with Skeeter Malcolm at the federal prison."

"Are you shittin' me?" Hank spat out.

"Did they set it up?" I asked, my heart thumping wildly.

"Skeeter's attorney did. The meeting was today at 3:30."

The knowledge that she'd gotten away from Wyatt and hadn't told me hurt me to my core. But I had to put my hurt feelings aside and focus on Carly's safety.

"She saw Malcolm?" I asked, my voice tight. "Did he tell her anything?"

"He did, but she refused to tell me what."

"How do you know any of this?" Hank asked.

"I was there at the prison," he said. "I didn't go in with her. Carter Hale—Skeeter's attorney—took her in as his assistant. I waited outside in the parking lot with a guy named Wyatt."

"*Wyatt?*" Hank and I practically shouted at the same time.

"I take it you know him?" Jed asked. "There's a warrant out for his arrest for kidnapping Carly."

"You knew that and let her *go with him?*" Hank demanded.

"Like I said, I tried to convince her to come with me, but she said she knew what she was doing and left with him."

"Jesus Christ," Hank said in disgust. "What is that girl thinkin'?"

"I don't think she was coerced," Jed said. "I got the impression they were working together."

"He kidnapped her at gunpoint," I protested. "Twice. That's not working together."

"Well, he wasn't holding a gun on her when they showed up," Jed said defensively. "I gave her multiple chances to get away from him, but she told me she knew what she was doing."

"Maybe he threatened her somehow," I said. "Maybe she was scared of him—"

"She sure as hell wasn't scared of him," Jed said. "She held her own when he got mouthy. I don't know why, but she chose to go with him."

I sat down on the bed, feeling numb.

Carly *chose* to be with Wyatt. Over me.

"He must have convinced her that he has something to help her with her father," Hank said as he sat on the bed next to mine. He leaned his crutches against the edge. "Something we can't give her."

"What the hell could that be?" I asked.

"She told me that you couldn't be involved, Marco," Jed said. "That she's asked you to cross too many lines."

That was like a punch in the gut. She really didn't want me with her.

"*I've* crossed plenty of lines," Hank shouted. "Why the hell didn't she ask for me?"

"She probably thinks you're still in Tennessee," I said, close to breaking down. We were all silent for several seconds before I said, "So she doesn't want me with her, and she thinks Hank's in Tennessee, but why Wyatt?"

"Because he's tricky, devious, and underhanded," Hank grunted. "Everything you need to deal with a weasel like her father."

"I think he's right," Jed said. "That's the only thing that makes sense."

"But he can use all of that against her," I said, dismay washing over me. "He's done it before."

"That boy's a smooth talker," Hank said in disgust. "He smooth talked me and I can usually spot 'em from a mile away."

"So she's falling for his bullshit?" I asked, not liking the possible implications.

"No," Jed said. "I don't think she's falling for it. Like I said, she was calling him out." He paused. "If it makes you feel any better, I gave her an iPhone."

I drew in a sharp breath. "Does it have a tracking app?" Hope filled me. "We can find her!"

Jed remained silent.

"You didn't put a damn tracking device on it?" Hank asked, sounding pissed. "What kind of enforcer are you, anyway?"

"It has a tracking device," Jed said flatly.

"Then tell us her location so we can go get her," Hank said.

But I already knew what Jed was going to say next, and after a few seconds he proved me right. "I'm sorry, but I can't do that."

"Can't," Hank growled. "Or *won't?*"

"Won't," Jed said firmly. "If Carly wanted you to know where she was, she would have let you know."

"She was kidnapped!" I shouted.

"As I said, she left of her own free will. One of my conditions for letting her leave was that she'd check in with me several times a day, and she's followed through. In fact, I spoke to her about ten minutes ago when she picked up a rental car."

"She got a rental car?" I asked in disbelief. This confirmed that she really wasn't being held against her will anymore. She'd chosen to do this with Wyatt.

"She was in a stolen van," Jed said. "It wasn't safe, so I got her something that was." He groaned, then said, "Look, she was intent on going to Dallas with that asshole, and I couldn't stop her. The only thing I could do was give her everything I could to protect her."

"And that's what we want too, you moron," Hank snapped. "We don't want to stop her. We want to help her."

"But for whatever reason, she doesn't want you to help her," Jed said. "She's a grown woman. I have to respect her decision."

"If anything happens to her, I'm comin' for you," Hank ground out through his clamped jaw.

"I'm not letting anything happen to her," Jed said as though Hank was being ridiculous. "I'm getting some backup, and then I'm following her. I'll let her do her thing, but at the first sign of trouble, I'll already be in position to help."

"We're already in Dallas," I said, running a hand over my head in frustration. "We can get to her faster."

"Well, she's not there yet," Jed said. "She's still in Arkansas."

"So tell us when she's in Dallas, and we can talk to her ourselves," Hank said. "Because I don't know who the fuck you are, and I don't trust you."

"I understand why you're upset," Jed said, "but if Carly wants you to know where she is, she'll tell you." He paused. "Marco, she asked me to tell you that she loves you."

I felt like I was five steps behind in this conversation, and my sleep-deprived, stunned brain was stumbling to keep up.

He continued, "And the next time I talk to her, I'll tell her that I've spoken to both of you. I'll tell her you to want to talk to her. That's the best I can do."

"Fuck you," Hank said.

"Did Carly give you my number?" I asked.

"No. After the Hardshaw Group bust, Carly became a little more free with the information she sent Rose in her emails. While she didn't give us *much* information, it was enough for me to track down your number."

I wasn't sure whether to be impressed or disturbed that he'd found my private cell number so easily, but if he could do that, he might have the resources to help me with something else. "If you can't give us Carly's number or location, then there's something else you can do. We're hoping to track down one of Blakely's old security men: Duncan Morris. He worked for Blakely when Carly's mother was murdered, but he quit soon after. Carly says Morris was kind to her when she was a girl, and she thinks he might be sympathetic enough to tell her what he knows. If I can't find Carly, then I can at least try to talk to him."

"How do you know he'll talk to you if Carly's not with you?" Jed asked.

"Maybe dangle that as a carrot the next time you talk to her," Hank snapped. "Tell her we're in Dallas, and we're goin' to see him with or without her."

I rubbed my eyes, trying to scrub away my exhaustion. "No, don't tell her that. Just tell her that we're in Dallas, and I love her, and to please let me help her with this." My voice broke. "And she shouldn't trust Wyatt. He's going to turn on her the first chance he gets."

"I'll tell her," Jed said softly. "I'll try to convince her to call you." He was silent for a moment, then said, "She's a lot like my Neely Kate. They were both treated like dog shit by their families, betrayed by men they thought loved them. Neely Kate felt like she had to do everything on her own until I convinced her we were a team. You just need to convince Carly that you're a team too."

"I thought she already knew that."

"Deep down she does, trust me," he said. "She just needs reminding."

Hank snorted in disgust. "If you two are done braiding each other's hair, then let's wrap up this call so we can get some real work done."

"I'll see what I can dig up on Duncan Morris," Jed said. "And I'll get back to you as soon as I find something." He hesitated, then his voice grew softer. "I can appreciate how frustrating this is for you,

Marco. If our roles were reversed, and you had information on Neely Kate and were withholding it—"

"She trusts you, Jed," I said. "Implicitly. I have no choice but to trust you too."

"Well, I sure as hell don't," Hank said. "And if you know what's good for you, you'll start telling us what we need to know."

"Here's what I *can* tell you," Jed said. "I'll let you know every time I hear from her. And I'll tell you when I get to Dallas. And when she gets to Dallas too. If you'll just trust me, we'll make sure she's okay, because that's what we all want, right? For her to be okay."

"Well, fuck you, razorback," Hank spat, "because I don't trust you one iota."

"I understand," Jed admitted. "But you have to know that I care about Carly. I wouldn't have gone to so much trouble to help her leave Henryetta if I didn't. I'm going to do everything in my power to help her. That's all I can give you for now." Then he hung up.

CHAPTER TWENTY-SEVEN

MARCO

"Who the hell does that fuckin' Jed think he is?" Hank demanded.

"He's trying to protect her," I said. "We have to respect that."

"Why aren't you pissed?"

"I *am* pissed!" I shouted, leaning over my legs. Truthfully, though, I was more frustrated than pissed. I had to respect Jed for standing firm on Carly's wishes, but I hated that she thought she had to protect me from this when she was the one who needed protecting. At least I knew she'd been okay this afternoon. Was she still okay, or had things switched back around after she left the prison? Sure, Jed had spoken to her when she'd picked up a car in Little Rock, but that didn't mean she hadn't been under duress. Wyatt could be calling the shots again.

Maybe I had to believe that because I struggled to accept she was with him by choice.

Then again, this wouldn't be the first time she'd teamed up with someone of questionable character. She'd had plenty of dealings with Bingham, and she'd always held her own. She could do the same with Wyatt. I had to believe that.

But why had she chosen to stay with him? What was her plan?

I wasn't getting answers any time soon, so getting information about Duncan Morris from Jed seemed about all I could do. Jed had worked for Skeeter Malcolm, a thought that should have horrified me but instead filled me with a mixture of relief and hope. He was used to getting things done. I had no doubts that he could dig up dirt faster than we could.

Hopefully, he could also convince Carly to contact me.

I glanced up at Hank, who was looking rough as shit. Worse than he had ten minutes ago. All this aggravation couldn't be good for him, and I had no idea when he'd last had a meal that hadn't come fried.

"Let's order room service," I said as I sat upright.

"Have you lost your fool mind?" Hank asked. "Or has staying in a place like this made you soft?"

"No to both." I got up and looked for a room service menu. "We need a decent meal and sleep."

"We need to protect Carly," he said, pacing the room with his crutches. "I should just drive over to her father's house and be done with this mess."

I stopped to look at him. "What does that mean?"

"Exactly what you think it means. I'm gonna kill the bastard."

"All the more reason to talk to Duncan Morris first," I said. "Find out what kind of security Blakely keeps around him, because I doubt we'll be able to drive over there, walk through his front door, and get immediate access to him."

"First of all, there's no *we* in this," Hank said through gritted teeth. "It will be *me*."

I wasn't about to let that happen but arguing with him would be pointless. "Well, you're in no shape to go killing anyone right now. You need a protein-heavy meal and sleep. You can go kill Blakely tomorrow."

Hank stared at me like I was an imbecile. "You think I'm gonna fall for that bullshit for one minute? There's no way I believe you're gonna let me go kill him."

"Don't be so sure about that," I said bitterly. "I'll do whatever needs to be done to protect Carly." If I were honest with myself, I believed the only way she'd be free from her father was if he were dead. But was I really willing to be part of his cold-blooded murder?

If it came to it, yes. I'd do anything in my power to save her.

My heart tightened, and I drew in a breath to relieve the pressure. I found the menu and handed it to Hank, but he just swatted it away with a shaky hand. Dark circles underscored his eyes, and his skin was pale.

"Then I'll order for you," I said, even more worried than before. "Maybe you should check your blood sugar."

"I can take care of my own damn self."

Only he hadn't been taking care of himself. Not until Carly had dropped into his life and taken charge. He was one more stubborn asshole she'd gotten to cooperate with her.

It made me feel a little better about her being with Wyatt. If she could tame Hank, surely she could deal with Wyatt too.

While I called in our order to room service, Hank grabbed the bag with the computer and opened it up. "I can do a little searching of my own on Duncan Morris," he said. "I'm not totally relying on that asshole."

"That asshole is helping her," I said, lying down on the bed, my shoes still on. I grabbed a pillow and stuffed it under my knee.

"Your damn leg still botherin' you?" Hank asked, his tone softer.

"Yeah."

"When was the last time you took something for it?"

"The last time we stopped. I'll take more ibuprofen when the food arrives."

His gaze narrowed on me. "You feelin' up to the task?"

"Of taking a couple of pills?" I asked sarcastically.

"You know damn well what I mean."

"I could ask the same of you."

He sat back on his bed, propped up against the pillows. "The sooner we get this done, the better."

I didn't like the implications of that.

"How's your insulin work? Carly has you test your sugar before you eat, right? Shouldn't you be doin' that?"

His face hardened. "You let me handle my own body, and I'll let you handle yours."

"That's the thing, Hank," I said in frustration. "I don't think you're handling yours at all, and if you let yourself get sick, you won't be at the top of your game when Carly needs you."

"I'll be fine," he grunted. "You take care of that leg. Who knows if we'll have to run."

I laughed.

"What's so funny?" he demanded.

"Can you see the both of us running from anything?" I said through my laughter. "Me with my limp and you on your crutches?"

"You're punch drunk," he said in semi-disgust. "You need some sleep. And I can move quick when I want to."

"I know you can," I said, closing my eyes. "I've seen you. I suspect you can probably move faster than me right now."

"Damn straight," he said. "Now rest your eyes while we wait for our supper."

"Yeah, good idea," I muttered as exhaustion made my eyelids heavy, then I drifted off into sleep.

My ringing phone startled me awake.

There was a room service tray on the desk, and Hank was napping on his bed, his laptop open on his lap.

I flailed around on the bed for my phone, then answered it, surprised Hank still hadn't woken up. It was Jed's number.

"Whatcha got, Jed?" I asked.

"We found some information on Duncan Morris. There's gonna be more, but I figured I'd give you the preliminary stuff."

I glanced at the clock on the table between the beds. I'd been asleep almost two hours. "Yeah," I said, scrubbing my face to help me wake up. "Tell me what you found."

"Morris worked for Blakely for about fifteen years before he left the company. His title was head of security. He resigned from his position about nine months after the death of Randall Blakely's wife and opened his own security business. Morris Protection Services. He provides security details to people of importance."

"But not Blakely," I said.

"Maybe he cut his teeth on Blakely, then decided to move on to bigger and better things," Jed said.

"Yeah, right."

He chuckled. "I didn't think so either. While Morris's security details fly around the world offering protective services, Morris tends to stick around Dallas. In fact, I know he's at an event tonight in Plano. Some fundraiser."

"Why would they need a security detail?" I asked.

"That's not why he's there. You got your badge on you?"

My chest tightened. "No. I resigned this morning."

"Because of Carly?"

"Partially." I didn't want Carly to feel guilty about my decision, and there was no denying my leg had something to do with it.

"Good," he said. "Can't have you waffling over your badge in this."

"Look," I said in a whisper-shout so I wouldn't wake up Hank. "Just because I'm no longer a deputy, it doesn't mean I no longer have any respect for the law."

"Maybe so, but by quitting you're acknowledging you may have to break it."

I ignored his statement, because we both knew it was true. "Now, are you going to tell me where this fundraiser is or not? Because once Carly reaches Dallas, I suspect shit's going to go down fast."

"My thoughts exactly," he said, a loud engine noise sounding in the background. "I was worried I wasn't going to get to Dallas in time, so we're flying. We'll be landing at a private airstrip about ten minutes from Blakely's house. I have a van waiting. After you meet with Morris, call me back. I probably won't have service, but leave me a message, then be ready to meet me at the airport. I'll send you the address."

"What?" I asked in shock. "Why the turnabout?"

"Because I kept putting myself in your shoes," he said. "Then when you told me you quit your job…well, that clinched it. You really are willing to do whatever you need to do to protect her, and I suspect we might need all the hands we can get to go against Blakely."

"Thank you."

"We've still got some time," he said, ignoring me. "They haven't even reached Texarkana yet. The highway's backed up. Looks like there was a wreck. I suspect it's bought us forty-five minutes, maybe an hour. I guesstimate she's about three hours from Blakely's house."

"You think she's going straight there?" I asked.

The sound of the engine grew louder. "I don't know, but I want to be prepared if she does."

"I think she will," I said. "There's no way she trusts Wyatt, so I suspect she'll want to get it over with."

"Neely Kate's doing some more digging. She'll call you if she has anything. I gave her your number."

"Thank you, Jed," I said, my voice breaking. "I mean it."

"Don't thank me yet," he said, his voice barely audible over the engine noise. "Neely Kate should have already sent you links to the files."

I knew he'd hung up from the silence on the other end. I checked my screen, and sure enough, another number had sent me links to a Dropbox folder with Carly's name. I signed in and found folders titled Duncan Morris, Randall Blakely, and even one with Jake's name. I opened the Morris one first. It had some biographical information—his birthdate, his home and business addresses and phone numbers, his dates of employment for Blakely, and information about his security business.

Then a text popped onto my screen from the number that had sent the files.

> Marco, this is Neely Kate. Duncan Morris is at the Silver Wing Fundraiser for Alzheimer's in Plano, but he's attending as a guest of honor, not providing security. I'll send you the address.

Neely Kate had been one of Carly's best friends in Henryetta. Maybe I could talk her around.

> Thank you for everything, Neely Kate. I can't let anything happen to her.

She sent back:

> That's our goal too, Marco. Rose and I are still digging for anything that might be useful.

> Can you give me her number? Please

My heart thumped hard in my chest as bubbles appeared on my screen and then disappeared. The bubbles appeared again, and a text popped up.

> I would give it to you if I could, but I don't even have it. Jed wouldn't give it to me either

I wasn't sure whether to believe her, but she had no reason to lie to me. She could have just said no.

> If you talk to her, PLEASE ask her to call me. I'm going crazy with worry.

> Jed says she's still in Arkansas. Get to Morris and see if you can get him to admit anything. We still have time to find her.

She was right. I needed to get to that fundraiser, but I wasn't exactly dressed for it, and neither was Hank. I glanced over at him, and he was sound asleep, snoring softly. The food service tray had two plates, one with a metal dome still over it. The other looked nearly empty. Since I'd requested two identical meals—chicken breasts with wild rice and broccoli—that meant Hank had eaten and was getting some rest.

Maybe I should leave him behind to sleep while I talked to Morris. But I needed to find a suit somewhere. A quick search showed a Men's Wearhouse about ten minutes away.

I glanced at my phone screen. It was a little after eight, which meant I needed to get moving, but since Wyatt had taken my bag, I was still wearing the jeans I'd put on the day before and a T-shirt I'd bought at a gas station in Memphis. Still, there wasn't time to buy a suit, then come back and shower and head out. The fundraiser started at eight and supposedly lasted until eleven. Would Morris stay there that long?

I decided to take a quick shower before heading to the store and took quick bites of my cold dinner as I got dressed in my old clothes. Hank was still asleep when I finished, something that actually concerned me, but his snoring was louder, so there was no doubt he was breathing. I wrote a quick note, letting him know that I was going to find Morris and I'd call him when I was done.

I slipped out the door, then jogged down the stairs to the first floor. It was only when I hit the parking lot that I realized this was an impossible task. The pants on suits off the rack weren't hemmed.

I didn't care. I wasn't going to let unhemmed pant legs stop me.

CHAPTER TWENTY-EIGHT

CARLY

Wyatt had been hostile until we'd picked up the rental car. But he seemed to perk up after I picked up the keys and checked in with Jed.

"I'll find somewhere to dump the van," he said. "You follow me."

Maybe it was a test.

I trailed him to a nearby, older neighborhood that looked semi-abandoned with lots of cars parked on the street. Everything in me screamed to take advantage of it and call Marco. To check in with him and let him know I was okay. But I worried I might lose my resolve if I heard his voice. And I just couldn't bring myself to drag him down with me.

I already knew there was a good chance this wouldn't end well. No sense in making him collateral damage.

Wyatt parked the van and sauntered over to our rental car, which I'd parked around the corner while I waited.

He opened the back passenger door, tossing my and Marco's bags onto the back seat, then shut the door and got in the passenger seat in the front.

"I'll nap on the way to Dallas," he said. "We've gotta be on our toes once we reach your father's territory."

"Yeah," I said absently.

Once we got onto the highway, I asked him nonchalantly, "So your deal with the FBI…how's that work?"

He'd been lying back in the passenger seat, but he was still awake. He glanced over at me. "Which one?"

"I get how the one in Drum worked. You were supposed to get evidence against your father."

"Not just *some* evidence. I needed to build a solid case to put him away for the rest of his life. They wanted me to build the whole fucking gallows," he said, getting aggravated. "And then everything I put together meant nothing to them, which makes no sense." He sat up, getting angrier. "So what if the state has a case against him? They can still prosecute him on federal charges based on the shit I've gathered."

"So they weren't interested in any of it anymore?"

"I don't know," he said with a sigh, sounding defeated. "I've been giving them information for months, so for all I know, they'll still use it. But they told me they don't need me anymore, and our deal is off." His expression tightened. "All the shit I did for them—visiting Louise in prison. Working with that scumbag Derek. If it weren't for them, Jerry might still be alive." His voice hitched. "I loved that man. I wanted to kill Derek with my bare hands," he said through clenched teeth. "Too bad his father got to him first."

"I don't know if I can ever forgive you for the role you played in getting Jerry killed," I said, my eyes stinging with tears.

"I know," he said, hanging his head. "I don't think I'll be able to forgive myself."

It all sounded shockingly similar to what Skeeter was dealing with, which made me glad I hadn't gone to the feds when I'd first fled Texas. They weren't to be trusted.

"So why go after my father?" I asked, swinging my head for a quick glance before turning back to the busy highway. His eyes were bloodshot, and he looked close to breaking down.

"Like I told you. It's my last chance at getting my deal back. If we can tie him to Hardshaw, they'll give me back my deal. Mom can stay in the house, and I can take over the businesses."

"And what about Max? What does he get?"

Wyatt's mouth twisted to the side. "I don't know. He's never shown an interest in any of my dad's businesses."

"Maybe because he didn't like working for your dad. He runs the motel for him and hates it."

"Maybe he'll want the motel," he said. "I'll ask him when this is done."

"If he's talking to you," I said wryly. "He's furious with you right now."

"I wonder why?" he sneered. "You turned him against me, Carly."

The fight bled out of me. We were talking in circles again. "You did that all on your own. And until you can accept that, you will never win him back."

He was silent for several seconds, then pushed out a long sigh. "Fuck, you're right."

I gave him a surprised look, and he released a wry chuckle. "Look, if there was ever a time for self-examination, it's now."

I sat up straighter and tightened my grip on the steering wheel. "So what are you going to do about it?"

His brow furrowed. "What do you mean?"

"It's not too late to save your relationship with Max. He loves you. He wants to be close to you, but all you've done is shut him out his whole life. Even when you two started talking again, you still held out on him. He's hurt."

He ran his hand over his head. "Yeah. I fucked everything up."

"So *call him*. Tell him you're sorry and you want to make it up to him. Tell him you'll stand by him to sort out this mess with your mother. He needs you, Wyatt, and you haven't been there for him."

"I know that too," he said softly, staring out the side window. "I've let a lot of people down."

"So, tell him you're sorry."

"You mean now?" he asked in disbelief.

"We don't know how this is going to go down. We might be walking into our deaths. This could be your only chance. Call him."

As I said the words, I realized that they were as true of me and Marco as they were of Wyatt and Max. If this went south, I'd never get the chance to hold Marco again. Or tell him that I loved him.

Wyatt sank back into his seat, his body slumping in defeat. "Yeah, you're right. I need to talk to him, if for no other reason than to give him closure. But not yet." He drew in a breath. "I'll call him when we get to Dallas. That'll give me time to come up with what to say."

I nodded because that made sense. I wasn't ready to call Marco yet either, and I'd want it to be in private. I'd wait until we checked into a hotel and call him then.

Still, I wasn't so sure I believed Wyatt. Knowing him, he might not intend to ever make that call.

We continued on I-40. Wyatt napped until we were about forty-five minutes east of Texarkana, before traffic slowed to practically a crawl.

I slowed down, which roused Wyatt, who took one look at the line of cars ahead of us then muttered, "Shit. There must be an accident ahead."

I grabbed my phone from the seat. Sure enough, there was a long red line on the map from where we were until Texarkana, ending in an accident symbol. "Looks like you're right."

He grabbed the phone from me, and when I gave him a warning look, he said, "I'm not taking it. I'm looking for a detour."

I let him have it because I was too impatient to sit in this line. The closer we got to Dallas, the more anxious I was becoming. I still believed I needed to confront my father, but I was nervous as hell about doing it.

"There's an exit about a half-mile ahead," Wyatt said, looking up and pointing out the windshield. "If we take that, we can cut across and pick up the I-30 in Texas."

"Sounds good. Let's do it."

By the time we got back onto I-30, we'd only lost fifteen minutes to the traffic jam.

"We need to figure out where we're going to stay," I said. "Jake and my dad live about a half-hour apart. I think we should stay somewhere in between."

"You know Dallas better than I do," Wyatt said. "I've only been here a few times. You lived here."

"When did you visit Dallas?" I asked.

He shrugged. "At least ten years ago. Once I went to a Cowboys game and the other time was with Heather."

Heather, his deceased ex-girlfriend. But they'd dated for years, so it made sense they would have traveled together. "Who'd you see the football game with?" I grinned. "Are you surprised I know the Cowboys are a football team?"

He laughed good-naturedly. "Nah. You worked enough Monday Night Football nights at the tavern that there would be something wrong with you if you didn't know." He turned serious. "I went with my dad. But not Max," he said bitterly, resting his arm on the console. "Just one more thing poor Maxy was left out of."

"You can still fix things with your brother," I said, shooting him a glance. "It's not too late. And when this is done, and you get to keep your family's land and businesses, you and Max can run them together."

"Yeah," he said, glancing down and then up at me. "You're right."

"I'm sorry your father pitted you and Max against each other."

He shook his head. "He did, but the truth is I knew my father shunned Max. So, to stay in his good graces, I did too." He took a shaky breath. "I'm not proud of it, but I need to own up to my actions. I chose my father over Max."

"We all screw up," I said, "and yeah, it may take Max some time to work through it, but if you keep showing him you're there for him, eventually he'll start to believe it."

He nodded. "I'm going to earn back his trust." He gave me a hesitant smile. "And everyone else's too." Releasing a chuckle, he added, "Although I'm not sure I'll ever regain Hank's trust."

I could have argued the point, but I didn't want to give him false hope. "Yeah, well..."

"I totally deserve his hate," he said softly, leaning his head back against the seat. "I screwed that boy over." Then, to my surprise, his body shook with a sob.

I could have told him it was okay, but it wasn't. He'd used Seth for his own gain, and the boy had been killed. He *was* partially to blame. The same could be said about the situation with Ruth and Franklin. "I can't tell you how to bear the weight of something like that," I said softly. "But I think the important thing is to own your mistake and try to make the world a better place to make up for it."

"Karma?" he half-laughed through his tears. "Don't you think I'm totally fucked if I'm counting on karma?"

I gave him a weak smile. "A lot of people who have done bad things have turned their lives around and become good people. You can do that too, Wyatt."

"Name one bad person who's managed that," he challenged with a laugh as he wiped his face with the back of his hand.

"Well...no one comes to mind right now—"

He laughed harder.

"—but there *are* people who have. And besides, there's only one Wyatt Drummond. Only you can decide whether you can change and to what degree."

He was quiet for a moment. "You think Max will forgive me?"

"I know he will," I said a little too eagerly, but I couldn't let him slide backward. The possibility of having a relationship with his brother might be the only thing keeping him from turning completely to the dark side. He had to have a reason to want to be a better person. Maybe Max was his reason. "Max has *always* wanted a relationship with you, Wyatt. But he might need some time to deal with...everything."

"He's goin' to blame me for Ruth's death."

I drew in a deep breath, then pushed it out. "Maybe."

His body stiffened.

"But I think he can move past it. Like I said, he just needs—"

"Time."

"Exactly."

"I've really fucked things up," Wyatt said, his voice breaking again. "Screwed up so many people's lives. Max. Jerry. Ruth. Franklin. Seth. Hank. Heather." He looked at me with sad eyes. "You."

When he said it like that, it really put things into perspective. It might take more than one lifetime to expunge his soul of all that bad karma.

"You haven't screwed up *my* life," I said, because I could only really speak for myself. "My father did a good job of that all on his own."

"But I lied to you and misled you and..." Sadness filled his eyes. "Lost you." He laughed again, bitterly this time. "Fuck, Carly, I kidnapped you at gunpoint. Twice."

I shrugged and gave him a sheepish grin. "Yeah, there is *that*..."

"Why are you here with me?" he asked incredulously. "Why the hell didn't you go with Jed?"

This conversion was starting to make me uncomfortable. "I already told you. Jed has a family."

"So why not ditch me?" He held up a hand. "And don't pull the *you need me* bullshit. It's more than that, and you know it."

"Yeah," I said quietly. "You're right. I've been thinking about it ever since we left the prison. I think, subconsciously, we both need this."

He shook his head. "What the hell does that mean?"

I made a face. "You're going to think I'm being cheesy."

A sideways grin spread across his face. "I'd expect nothing less from you."

A sense of mourning washed over me because he reminded me so much of the Wyatt who befriended me when I'd first come to

Drum. For a brief period, I'd thought the two of us could conquer the world, or at least both of our fathers.

"We made a promise to each other. First your father, then mine," I reminded him solemnly. "We need to finish what we started." As stupid as it seemed, part of me truly believed that.

His smile fell and he looked surprised.

"Plus," I added. While part of me wanted to believe we could do this together, I hadn't forgotten that Wyatt had betrayed me and lied to me at every turn. I wanted to test his reaction to what I was about to say next. "This is a way to prove to Max that you've changed."

He considered me for a moment. "You're right. We need to finish what we started. I made you a promise, then broke it."

"Of course, it's not lost on me that you're getting something out of it," I added dryly.

"You got something out of my father's arrest," Wyatt pointed out. When I lifted my brow in a sarcastic look, he added, "You gained the confidence to take down your dad."

"And knowing I'd helped protect the town from a murdering, blackmailing tyrant."

He shrugged. "There is that."

Was he throwing my own words back at me on purpose or was he parroting what he thought I wanted to hear?

I had to admit that all his responses seemed genuine. His remorse, his tears, his grief.

"I know what you're thinking," Wyatt said, studying me. "You're still wondering if you can trust me, which is fair. I sure as hell wouldn't trust me. What can I do to prove I'm on your side?"

"I don't know." I was stupid to consider it a possibility, and yet here I was, doing just that.

He nodded. "We still have an hour until we get to Dallas. If you think of something between now and then, let me know."

Truth serum? Telepathy? But I doubted I could get either of those things before reaching Dallas.

CHAPTER TWENTY-NINE

MARCO

A half-hour later, I was dressed in a black suit, black silk shirt, and a black silk tie, along with black dress socks and shoes. The salesclerk had used double-sided tape to hold up the hem of my pants after I'd paid her fifty dollars to help me out.

The event was being held at a hotel forty minutes away, which meant I'd be arriving at around nine-thirty. I suspected the affair wouldn't be in full swing for at least an hour. Hopefully, Morris would still be there.

The parking lot was full when I arrived, and I had to park in the back lot. The night was warm and humid, and I started to get hot as I headed toward the hotel. I hurried toward a back door as a couple came out, and thankfully they held it open for me. The woman glanced back at me and smiled, then turned to her scowling partner.

Once I was inside, I pulled out my phone and opened the file on Morris again. Neely Kate had included multiple photos of the tall, silver-haired man. He was in his early sixties and looked like he could still kick someone's ass with his hands tied behind his back. I suspected he wouldn't be hard to spot.

Okay. Now all I had to do was convince him to tell a total stranger that he had proof Randall Blakely had murdered his wife.

I was going to get *my* ass kicked.

But I was still going to talk to him.

I made my way down the hall toward the front of the hotel, then followed the sound of music. I wasn't sure how I was going to get past security, but there weren't any gate keepers posted at the doors. Then again, once I got inside, it looked like the dinner was nearly over. The guest of honor was giving his speech on a small stage at the front of the room.

I hung in the back of the room as Duncan Morris talked about his mother's struggle with Alzheimer's. At the end of his speech, he received a standing ovation as he picked up a foot-tall crystal award. Multiple people had lined up at the edge of the stage to greet and thank him.

The service staff began to clear tables as a woman announced that the party was just getting started and a DJ tried coaxing people out onto the dance floor.

A woman in a formal dress made her way to me and gave me a warm smile. "You look a little lost." She swung her long blond hair over her shoulder with a flirty look.

I smiled back. I wasn't about to get kicked out now. "My date is talking to her friends, and I needed to stretch my legs."

Her brow lifted. "And who's your date?"

Shit. Was she trying to bust me or pick me up?

I gave her a sheepish look. "How badly will you judge me if I admit that I forgot?" I cringed. "It was a last-minute thing. Her first choice canceled on her, and a friend of a friend asked me a couple of hours ago." I shrugged. "My grandma has Alzheimer's, so I figured, why not?"

She eyed me up and down. "I didn't notice you earlier. I'm surprised I missed you." The look on her face suggested she was happy she'd noticed me now.

"Say," I said, struggling hard to play it coy. I felt like I'd lost my game. Or maybe I just didn't need or want it anymore. "My date looks like she'll be busy for a bit. Would you like to dance?"

A smile lit up her face. "I thought you'd never ask."

I darted another glance toward Morris as I placed a hand on the small of her back and ushered her to the dance floor. I'd done this with countless women, but it felt wrong now, dirty. Nevertheless, I couldn't stand against the back wall like a creeper, and Morris was still surrounded by his many admirers, most of whom were mature women. No surprise since his dossier had said he was single.

When we reached the dance floor, only a few other couples were dancing to a pop song. My partner placed her hand on my chest. "What kind of moves do you have?"

I grinned. "A few." I started to dance, which didn't come as easily as it had in college, when Max and I used to go out every weekend, but I held my own despite the cramp in my thigh.

It didn't take game to know she was trying to pick me up, and a year ago, I wouldn't have hesitated. Now, the thought of sleeping with anyone other than Carly made me physically nauseous.

It took another two songs before the crowd around Morris started to thin out. I excused myself, saying I needed to check on my date, hoping she didn't pay too much attention to where I went.

I headed over to a group of people closer to Morris. Sure enough, the woman's gaze followed, but she moved on after a few seconds. After making some casual small talk, I excused myself to go to the bar and ordered two whiskeys. I was spending money like it was water tonight, but I'd spend it all if it meant saving Carly.

After the bartender poured the drinks, I headed up toward the stage, where Morris was talking to one last straggler.

I stood behind her as she told him how much she admired him—all while touching his arm and offering to work with him personally if he wished.

He gave her a warm smile. "Thank you, Wendi. I will definitely keep that in mind. Be sure to email me your contact information."

She blushed, then promised she would before walking away.

Morris's gaze landed on me, and he gave me a questioning look as he eyed the drinks in my hand. "You're either a life saver or a lush." He looked me up and down. "And you don't look like a lush."

I handed him one of the glasses. "Good observation, but then again, I would have expected as much."

Taking the glass, he lifted it to his mouth. "I also noticed you came in toward the end of my speech." He took a sip, then said, "Great timing. You didn't miss much."

I was surprised he'd noticed, but then again, he was the head of a security agency.

"I take it you're here to see me," he said, taking another sip. "Are you trying to convince me to take a job?"

"In a way..."

He nodded slightly. "I admire your determination. It's a little loud in here. Why don't we walk over to the hotel bar?"

"You're not needed here?"

"Not at the moment."

We walked toward the back doors together. Morris downed the rest of his whiskey and set the glass on a table before we walked out the doors. I set mine down next to it, untouched.

Morris walked toward the bar with purpose, then sat at a high top for two and flagged down a waitress. "I'll take two Woodford's neat." He gave me a smile as she walked off. "Much better whiskey in the bar." Sitting back on his high-back stool, he asked, "We're not really taking new clients, but I admire your ingenuity. Tell me about the job."

I hoped he still admired my ingenuity in about thirty seconds.

"I'm going to give it to you straight, Mr. Morris."

He held up a hand with a practiced wave and smile. "Call me Duncan. Please. And I prefer it straight, Mr....?"

"Roland," I said, extending a hand. "Marco Roland."

He shook my hand, then sat back again. "What type of security are you looking for? An event? A bodyguard?"

"More like protection," I said. "For my girlfriend."

His brow lifted, but it seemed like a practiced movement,

something he'd worked on as part of his spiel to make all his clients feel special. "Does she have a stalker? Have you contacted the police?"

"She's afraid to contact the police. The person who's a threat to her has a lot of power, and she knows they'll take his side. She's unprotected and he's moving in. We don't know how much time she has left before he finds her and kills her."

He looked genuinely concerned. "And you're sure you don't want to involve the police?"

"No. We feel like it's best handled privately."

His eyes brightened. He was on the hook. "One of them is a person in the public eye."

"Yes."

The waitress returned with our drinks and set them on the table. "Let me know if you need anything else," she said to Morris, then offered me a lingering look. "Anything at all."

I barely gave her a glance before focusing my attention on Morris.

He grinned. "You must really love your girlfriend."

"I do. Her name is Carly." I slipped my phone out of my pants pocket and opened my photo app, pulling up a photo of Carly. She was sitting on my porch, wearing a blue sundress that brought out the blue in her eyes. Smiling at me with such unabashed love it stole my breath every time I looked at it.

I handed it to him. "Here's a photo of her."

He gave it a cursory look, then stopped, a strange expression washing over his face. Like he'd seen a ghost. "What did you say her name was?"

"Her nickname is Carly," I said. "But her legal name is Caroline."

His gaze lifted to mine, his eyes wide. For a moment, I thought he was going to bolt and leave me with nothing but the tab for the expensive whiskeys.

"Caroline?" he finally choked out as he put my phone on the table and pushed it toward me. "How long has she been in danger?"

"Since last August." I held his gaze. "She ran away the night before her wedding. She's been in hiding ever since."

He stared at me for a hard second, then downed his whiskey like a shot. "Fuck," he grunted as he set the glass down, his previously affable expression replaced by dismay. "Where's she been hiding?"

Did it matter now? "Eastern Tennessee. After a couple of months in Arkansas."

He nodded, then flagged down the waitress. "I need another." He turned back to me. "You say you're her boyfriend? How's that work? The news said she left a devastated fiancé behind."

"Seems to me you should be focusing on the bigger issues, like who threatened to kill her," I said dully. "I don't think there's any point in pretending you don't know who I'm talking about."

He drew in a breath, then pressed a finger on the table as he considered my words. "How do I know you're really with this woman?"

I picked my phone up from the table and woke up the screen, then pulled up a selfie of Carly and me I'd taken that same afternoon. "She dyed her hair dark to change her look and changed her name. She's been hiding in a small Appalachian town, which is where I met her, to be clear. She wasn't cheating on her fiancé," I said in disgust. "Drum has a population of about two thousand, with little internet access and no CCTV. It was easy for her to keep from having any kind of digital footprint. But she didn't want to be looking over her shoulder for the rest of her life, wondering when she'd be found."

"She's stirred the pot," he mumbled as he started scrolling through my photos.

"Yes." I could have stopped him, but I had nothing to hide. If looking at photos of Carly persuaded him to help us, I'd send him copies. I'd even frame them.

"She looks so much like her mother," he said as he scrolled. "When I saw the reports about her wedding, I feared the worst.

Then Blakely went public a couple months later, and I knew he must be desperate. But I hoped that it meant she was still alive."

"She's alive, but we're not sure for how long. And he didn't go public because he was quietly searching for her. We think he was close to finding her in Arkansas, but she caught wind and fled again, with a new identity. A week or so after that, they announced she was missing."

He nodded. "He flushed her out and then cast a net to catch her."

"She was in Drum by then. No one there made the connection. Hell, hardly anyone even noticed the news report. Besides, she looked different from the pictures and didn't act like she came from money. We all just saw her as Carly, the sweet new waitress at the tavern."

"She was sweet when she was a girl," Morris said, handing me back my phone. "I loved seeing her when she came to the office." His expression became more stern. "Why exactly have you come to me, Marco?"

"Because I think you can help her," I said.

"Providing security? I don't come cheap."

I wanted to throw my drink in his face. "You'd charge her for protection? When you helped the man who's trying to kill her. *Really?*"

His eyes flew wide. "Keep your voice down!"

I leaned closer and whisper-hissed, "You don't want your many admirers to hear that you murdered a woman in an accident that was supposed to kill her eight-year-old daughter as well?"

"I had no part in that accident," he said, his face turning red. "None."

"You're not denying it was murder."

He slammed down his third drink, thunking his glass on the table. "Fuck you."

I didn't see a point in responding.

"What exactly do you want from me?" he asked with a haunted look.

"I need proof of what he did."

Morris shook his head. "I don't have it. Trust me, if I did, I would have turned it over to the police the minute I had it in my hands." He propped his elbow on the table, resting his forehead in his hand. "But I knew deep in my gut he'd killed her, and I couldn't be a part of an organization that condoned that. So I resigned less than a year later."

"And what about Carly?" I asked, my anger rising. "I just fucking love how everyone washed their hands of him and conveniently forget they left *an innocent child* behind."

"He loved that girl," Morris said, jabbing a finger at the table.

"If he loved her so much, then why do you think she ran from him? He was plotting to kill her before her wedding. He wanted her asshole fiancé to be part of his criminal organization, but they had arcane rules. Only family could be involved."

He sat back, his guilt written all over his face. "Fucking Hardshaw Group," he muttered.

"So you knew about them?"

His eyes hardened. "I was his head of security. Kind of hard not to know."

"So you left her in that mess too?"

"*She* didn't know about it," Morris said like I was stupid. "And as I said, I thought he loved her."

"Apparently he did." I gave him a bitter smile. "Until he didn't."

He placed his hands on the table and spread them out. "I had my suspicions about the accident but couldn't find any proof. Believe me, I tried."

"So who sabotaged Carly's mother's car?"

He shrugged. "It had to be someone in Hardshaw. Blakely must have thought everyone would pass it off as an accident. But then I remembered how Mary Caroline's parents were killed, and I looked it up. The similarities…they were too strong to be coincidental."

I sat back, feeling disheartened. "He's closing in on her."

Morris's eyes narrowed and he leaned closer, his face intense. "Then she needs to run."

"She doesn't want to run. She wants a real life. A family of her own. She can't do that if she's constantly looking over her shoulder."

"*You* want a family. You don't want to be looking over *your* shoulder," he sneered.

I really wanted to punch the smug look off this asshole's face. "I'd run with her to the end of the earth if that's what she wanted, but it's not. She's pissed as hell, and she plans to fight him."

"She'll never win."

"You don't know of any evidence that will put him away? Anything at all?"

"Randall Blakely is too damn careful for that."

That's what I was afraid of. "Okay. Let's say she wanted to confront him, where would be the best place to do that?"

His mouth dropped open and he started to laugh. "You can't be serious."

"Unfortunately, serious as a heart attack."

"You have to stop her," he said, sitting up taller. "That's insane."

"That part's a given," I said. "But for the sake of argument, let's say she chose to do it anyway. How should she go about it?"

"First of all, I'd tell her not to do it. But if she's insane enough to try, I'd tell her to do it at his house. He usually has less security there."

"How many guards?"

"I haven't done his security in years, but I'd guess two. Heavily armed. He'll also have tons of cameras. There won't be any way to sneak onto the property. They'll know within seconds."

Basically, we were fucked.

CHAPTER THIRTY

MARCO

"So you won't help her?" I asked, already knowing the answer.

"Not only no, but hell no," Morris said, shaking his head. "I was lucky to get away from that man the first time. I'm sure as hell not putting myself in his line of fire."

"You're a big security guy," I said. "The best of the best."

"I'm not going up against a crime syndicate. No way."

"But that was supposedly busted up," I said.

"Was it?" he asked, raising an eyebrow. "One can never be too sure." He stood and dug out his wallet, pulling out several twenties. "My advice? Run. Leave the country." He closed his wallet and stuffed it back in his pants. "Good luck to you, Marco. You're gonna need it." Then he headed back to the ballroom.

"Cowardly asshole," I grunted under my breath as I picked up my glass and took a sip. Then again, I would have been more surprised if he'd offered to help. Everyone in Carly's life had scurried away from the conflict between them like mice from the kitchen when the lights are turned on.

But *I* wasn't running away from him. I was running to her.

Sliding off the stool, I pulled out my phone and saw I'd missed a call from Hank. He'd left a voicemail that made it pretty clear he'd found my note and wanted me to call him immediately.

After I initiated a call to him, I started walking down the hall to the back door to the parking lot. He didn't waste any time lighting into me.

"What in the Sam Hill do you think you're doing?" he shouted in my ear. "We were supposed to see him together!"

"You were sleeping. I didn't want to wake you."

"You fell asleep too," he snapped. "And I didn't leave you to go murder Blakely!"

Most likely because he wouldn't have been able to stay awake long enough to drive there. "Do you want to know what I found out or not?"

That took the bluster out of his voice. "Morris let you into his house?"

"I didn't go to his house. He was the guest of honor at an Alzheimer's fundraiser dinner. I got here after he finished his speech, and surprisingly, he talked to me."

"I don't believe that," he scoffed.

"I was surprised too," I said, knowing he wasn't accusing me of lying, "but he thought I wanted to hire him, so I went with it and told him my girlfriend needed protection and showed him a photo of Carly. He instantly recognized her."

"What did he tell you?"

"Not much," I said. "But he confirmed that he thinks Blakely killed his wife and in-laws. He just never found proof. He quit because he couldn't stomach working for someone like that."

"So the visit was a wasted effort."

"I wouldn't say that," I said, moving closer to the wall as two women headed down the hall toward me, pulling large rolling suitcases behind them. I lowered my voice. "He said the best place to confront Blakely is at his house. He thinks Blakely will only have two security guards. Heavily armed, of course."

"That might be a challenge for us," Hank said.

I told him that Jed was flying in, and what he'd said about Carly being a little ways behind.

Hank was silent for a moment. "Maybe, maybe not. She's bound to be exhausted, and I suspect she'll want to tackle it when she feels more alert. She's gonna need to be on her toes."

"But she's with Wyatt," I pointed out as I walked out the door to the parking lot. "I know she thinks he's working with her, but I figure it's only a matter of time before he turns on her. He's going to want that reward money."

"You might be right," Hank said. "Plus he knows we're on his tail, and he has to know about the phone Jed gave Carly. Wyatt's not a stupid man. He's bound to presume it has a tracking device. Which means he's going to want to beat us all to Carly's father's house. So we need to get our asses over there ASAP."

"What if Wyatt's already made arrangements to hand her over? It might not happen there."

"And if he's prearranged it and it *is* Blakely's house, there's bound to be more than two guards," Hank said. "That makes things a hell of a lot trickier."

He was right. Dammit.

"I take it Jed's still in the air?" Hank said.

"Yeah." I stopped next to Hank's car and unlocked the door. The car was so old it used a key next to the door handle. "He said he'd call as soon as he landed."

"And he's the one with the tracking device."

My heart sank. "Yeah. His wife, Neely Kate, texted me information about the fundraiser I just crashed. I asked her if she could give me Carly's number, but she claimed Jed didn't give it to her."

"You think she's lyin'?" Hank asked.

I got in the car and started the engine. "No. What reason would she have to lie? I got the feeling she'd tell me if she had it and didn't want to give it to me."

"Always presume everyone is lyin' to you," Hank said. "Didn't they teach you that shit in deputy school?"

I couldn't help grinning as I backed out of the space. "Yeah, something like that."

"So we're at Jed's mercy."

I headed toward the parking lot exit. "Trust me, I don't like it any better than you do. But we've got no choice in the matter."

"That won't stop me from chewin' Carly's ass out when this is all said and done."

"She'd expect nothing less, Hank," I said as I turned onto the road. My phone beeped with an incoming call from Jed. "I'm getting a call from Jed," I told Hank quickly. "Let me take it, and I'll call you right back." I didn't wait for him to respond, just ended the call and accepted Jed's. "Marco."

"We're still in the air, approaching the airport, but low enough to get cell signal," Jed said, the loud engine sound muffling his voice.

"I talked to Morris, but I didn't get much out of him. He thinks Blakely's house will have two heavily armed guards. But Hank and I suspect Wyatt may have already made the arrangements to hand her over to her father for the reward. If so, he may have been instructed to take her somewhere else."

"That's why I called. There must have been a glitch in the tracking information, or they found a faster detour, because she's already in Dallas."

"*What?*"

"She's on her way to Blakely's house. In fact, she's only a few blocks away."

"Shit. I'm at least twenty minutes away." My heart pounded in my temple. I wasn't going to get there in time.

"I'm about the same distance." Jed sounded breathless. "We're at least five minutes from touching down, then we have to taxi and get to the car." He paused and cursed under his breath. "I've called and texted her multiple times, but she's not answering."

"Wyatt's taken control," I said, feeling like I was going to throw up. "He's going to hand her over to her father."

"Even if he does, Blakely won't kill her right away. In fact, I doubt he'll do it himself. Men like him always have someone else do their dirty work. He'll want to talk to her first. Play with her. Show her that he's the big bad boss. Then he'll probably have his men take her somewhere off his property to do it. We have a bit of time."

"If we can get past the guards."

"We'll get past the guards," he said confidently. "You head to Blakely, park down the street and wait for me. We'll go in together."

"I left Hank back at the hotel," I said with a groan. "And it will take too long to go get him."

"We'll be fine without him. We need to focus on getting to Carly ASAP."

"Agreed. See you there."

I hung up and called Hank back, now cursing myself for leaving him behind.

"Carly's already in town," I said when he answered. "She was a few blocks from Blakely's house."

"You just had to fuckin' leave me here!" he shouted. "Because now you don't have time to come get me, do you?"

"No," I admitted, regretting it with every fiber of my being. "I fucked up."

"I'll say you did. Do you even have a gun?"

Fuck. "No."

"There's a handgun in the glove compartment," he said, sounding out of breath. "Take it, and I'll meet you there."

"How will you get there?"

"I'll get there the same way everyone else in new-fangled world gets around when they don't have a car. I'll take one of those Uh-bers."

"You mean Uber?" I asked in shock.

"Don't sound so surprised. Now get your ass to Blakely's, and I'll meet you there soon."

Hank was about twenty minutes from Blakely's too. I merged onto the highway and drove fifteen miles over the speed limit, weaving in and out of traffic.

I drove like my life depended on it because I knew Carly's probably did.

CHAPTER THIRTY-ONE

CARLY

My nerves kicked in as soon as we crossed the city limits. Of course, we were still miles away from my father's house, Jake's house, or even my father's office, but being in the same metropolitan area as those places was enough to make me anxious. I could feel the scared little girl slipping back inside of me, terrified for her life.

I wasn't that girl anymore. I was a strong, confident woman.

But that didn't mean I could defeat my father.

"Take a deep breath, Carly," Wyatt said softly next to me, and I gave him a look of surprise. "You look like you're about to pee yourself."

Great. So much for being a strong, confident woman.

"Let's pull off at the next exit and take a moment to regroup," he said. "We've been in the car for most of the day. We could stand to stretch our legs."

I nodded, if for no other reason than because I needed to catch my breath. I could feel a panic attack brewing.

There was an exit ahead of us, so I pulled off and turned into the first gas station I found. As soon as I parked in front of a pump, I opened the door, fumbling to unlatch my seatbelt.

"Carly," Wyatt said, reaching for me, but I pulled out of his grasp, scrambling out of the car. My chest tightened, and I struggled to take a breath.

Wyatt was beside me in seconds, worry in his eyes. He rested his hands on my shoulders. "Carly. It's okay."

"I can't do this," I forced out through my tears. "I can't face him. I thought…I…could." My breaths were coming in short pants. "But… I…can't."

He gave me a sad smile. "You don't have to face him at all, Carly. Why don't we work the Jake angle? He doesn't seem like the sharpest tool in the shed. I bet we can find *something* incriminating in his house."

My mind was racing, arguing with itself, torn between the conviction that I'd never be free if I didn't do this and that I was insane to believe I could pull this off. That I could stand up to that man for the first time in my life.

"Deep breaths," he said, holding my gaze while showing me how I should breathe.

I fought the invisible band constricting my chest to drag a deeper breath into my lungs.

"That's it," he said encouragingly. "Again." We both breathed deeply together, my chest expanding more this time.

It took me nearly a minute to calm down. People gawked at the hyperventilating woman with tears streaking down her face, but Wyatt ignored them, keeping his focus on me. Once I'd calmed down, he cupped my cheek. "We don't have to do this at all, Carly. We can go back to Tennessee."

"What about your FBI deal?" I asked, suspicious that he'd let it go so easily.

"If Jake really has evidence in his house, then I won't need you," he said with a wry grin. "Just give me the security code, and I'll handle everything. We can use the evidence to put away your father. You don't have to face him. You can watch it all go down on the news." He tilted his head to the side. "In a way, it's more fitting, don't you think? You brought down my dad without me, and I can reciprocate and bring your dad down for you." He seemed to be warming up to the idea. "In fact, we can get in the car, and I'll drive you to the airport. You can use your phone to book a flight anywhere. Where do you want to go? Knoxville or Nashville if you want to go back to Tennessee, or maybe you should run again and have Marco meet you somewhere. We both know Jed will get you both new identities if you ask. You can start out fresh, but you'll be safer because your father will be in prison." His thumb brushed my cheek and his gaze softened. "Let me do this for you, Caroline. It's my turn to keep you safe."

I had to admit, I was tempted. It would have been so easy to let Wyatt do this for me, and if Jed had offered to do it last August, I would have taken him up on it in a heartbeat, then hidden in the basement of Rose's farmhouse until it was done. But I wasn't Caroline anymore. I was Carly Moore, and I'd never be able to live with myself if I let Wyatt or anyone else do my dirty work for me. Besides, if I ever wanted to free that scared little girl trapped inside of me, I needed to face my fears head on.

Which meant I *had* to face my father.

I took several deep breaths, still scared but controlling it now. Not letting it control me.

"No," I said, reaching up to swipe the tears off my face. "I have to do this, or I'll never be able to live with myself."

Sympathy filled in his eyes. "Carly..."

I took in another deep breath. "No. I'm not running this time. I'm facing him." But now that I'd made this decision, I didn't want to mess with Jake. Wyatt could investigate that on his own.

And I could deal with my father on *my* own.

Deep down, I think I'd known this was the way it was supposed to be. Maybe that's why I'd come with Wyatt and hadn't called Marco. Because Marco would never have let me do this on my own, and it would have killed me to leave him.

"I could really use a bottle of water," I said with a soft smile. "Do you think you could go in and get me one?"

"Of course," he said, rubbing my upper arms. "Do you want anything else? A snack? Coffee?"

"No," I said. "Just the water." I motioned to the pump. "I'll fill up with gas while you're in there."

"Okay." He reluctantly turned and headed inside while I leaned inside the car, pretending to get the debit card out of my purse, but as soon he was inside and out of sight, I got in and turned on the engine.

He was out of the door in flash. "Carly!"

I ignored him, instead backing up and then turning the wheel and pulling forward, heading toward the road. Wyatt ran behind me, shouting my name. I barely stopped before pulling out onto the street. Wyatt stood at the edge of the parking lot, his hands on top of his head in frustration.

I glanced down at my phone, surprised he hadn't called me yet to complain, but then I remembered he didn't have the number of the phone Jed had given me.

I really was on my own.

It felt surreal as I drove to my father's house, like I was watching a movie about someone else's life, and I was the actress playing myself. Then again, maybe I *had* lost my mind, and this was the only way I could cope with my own crazy decisions.

But I was still sane enough to know showing up at my father's house without a plan was pure idiocy. While confronting him might feel good for about thirty seconds, the bullet his security team would likely put in my forehead would hurt.

This was more than me confronting my fears. This was me getting justice for my mother. I planned to make him confess, and I needed that confession to be heard by other people. They needed to know that he was the monster I knew him to be.

But I knew the police probably wouldn't use a taped confession, even if I managed to get one and, miraculously, to get away, which seemed like a long shot. I suspected he'd greased enough wheels of justice to slide right out of any indictments.

No, it wasn't the police who needed to hear his confession. It was the whole world. Some people wouldn't believe it, but many more of them would. Then we'd see if he could get away with it.

When I was a few blocks away from his house, I parked at the curb and downloaded the Facebook app. It had been so long since I'd been on social media. I wasn't sure I'd remember my password as Caroline Blakely, but after the app finally downloaded, it only took two tries for it to open and show my hundred-plus notices.

I was in.

I ignored my feed and the notices, instead focusing on how to start a Facebook Live. It took a half-minute to get it set up and ready to go.

After I cued up the video, I slipped on the jacket I hadn't returned to Jed and put the phone in my pocket. Audio was good, but I'd prefer video too. Just so there wasn't any room for doubt.

I'd figure out how to get *that* once I was inside.

Stuffing the gun in my purse down into my jacket pocket, I got out of the car and walked the two blocks to the gated entrance to my father's house, then rang the bell. I'd considered trying to sneak in, but what was the point?

Go big or go home.

"Do you have an appointment?" an elderly male voice asked through the intercom. I recognized my father's butler's voice.

"Yes, Reginald," I said, my voice firm. Now that I'd resolved to do this, my fear had faded, replaced by determination. "Tell him Caroline is late to see him."

I heard his gasp, then the intercom line went dead. Seconds later, the doors swung open to the long drive leading to the house.

I started walking.

Sure, I may have lost the ultimate element of surprise, sneaking up on him while he sat in his study, sucking on a cigar, but I had no doubt his shock would still be on his face when I showed up in front of him after all this time, when we both knew why I'd run.

He'd have questions.

And so would I.

CHAPTER THIRTY-TWO

CARLY

Reginald was waiting at the front door when I walked up the steps. He held it open but partially blocked the entrance, fear etched on his pale, wrinkled face. He'd worked for my father since before I was born. I suspected he knew this wouldn't go well for me.

"Miss Caroline," he said in his raspy voice. "We've all been so worried."

I gave him a kind smile. He'd always been good to me. After my father fired my nanny, pretty much leaving me on my own, Reginald and Leta, the housekeeper, had tried their best to take over.

Ignoring protocol, I wrapped my arms around him and gave him a hug.

His body stiffened, then a second later, he stepped onto the porch and hugged me back. "You need to leave," he whispered in my ear. "Your father..."

I pulled back and looked into his eyes. "I know."

Shock covered his face. "I don't think you do."

Stretching up on my tiptoes, I kissed his cheek, then walked past him into the round, two-story entryway. "Is he in his study?"

"Y-yes," he said, following me inside. "But wouldn't you like to go..."

I ignored him, heading for the circular stairs. Nearly a year ago, I'd gone up these stairs, happy, content with my decision to marry my lifelong friend. After seeing all my friends' parents' messy marriages, I'd figured friendship was a better basis for marriage than chemistry and all-consuming love.

Turned out I was wrong about that.

Or maybe it wasn't fair to make that judgment based on Jake since he'd turned out to be the best friend from hell.

My father was probably watching me now. I knew he kept cameras focused on the entryway so he could see who was coming to get him if the boogeyman broke in. Lifting my chin, I continued up the fifteen-foot staircase, steeling myself for what was to come.

A hall extended from either side of the landing. I knew my father's study was to the right, so I headed down that way, the rich wool rug absorbing my footfalls. The corridor was dimly lit by wall sconces, giving the house the look and feel of a home built in the late 1800s.

I knew for a fact that it had been built in 1985.

This place had never felt like home. Hank's house, Marco's house, even the tavern felt far more like home than this place ever had.

I stopped next to a door. Between the camera angles and the lighting, I knew this was a blind spot, one where my father wouldn't be able to see me for a brief period. I pulled my phone out of my pocket, woke up the screen, and started the Facebook Live. I slipped it down into the middle of my bra, adjusting it so the camera lens peeked through a small keyhole gap in my dress at my breastbone. I prayed the fabric didn't end up blocking the lens and that my father's study was dim enough to hide it, but not so dim as to make the video too dark.

I took a deep breath and tried to steady myself. This was it.

Could I pull this off?

I didn't have many Facebook friends, so I doubted many people would initially view the feed. I'd made it public, hoping people would share it, but I ran the risk of someone my father knew seeing the video and then calling to alert him, which meant I'd need to act quickly before someone stopped me.

I headed the rest of the way down the hall, stopping outside his study, thinking again about last August when I'd stood outside Jake's old room in his parents' house, listening to the two of them scheme about my murder. I'd gone from naïve yet content to terrified. One minute was all it had taken for my life to turn on its head.

Would it take longer than a minute tonight?

If nothing else, I might capture my own murder on a live feed, and they would certainly put him away for that.

I knocked on the door but didn't bother waiting for him to answer like I'd learned to do after my mother died. Half of the time he'd never respond at all, leaving me to go back to my room at the other end of the house.

Eventually I'd stopped trying.

I flung open both doors, leaving them open behind me—a strict no-no when I was a kid.

I was breaking *all* the rules tonight.

My father was in his favorite Italian leather chair, just as I'd imagined him, and he had the cigar too. His face jerked up, and the sheer shock on it suggested it wasn't just me barging in that had caught him unaware. He hadn't known I was here.

I had a surprise of my own. Jake was sitting in the chair opposite him.

"Hello, Father," I said, walking in and stopping in front of them both.

"Where have you *been*, Caroline?" Jake asked incredulously as he got to his feet. "You abandoned me the night before our wedding without a single word, didn't let me know where you were, and now you've dropped in like you've just returned from the rapture, and you *dare* to ignore me?"

I gave him a bored expression. "Shut up, Jake. The grownups are talking."

His face contorted in outrage. "How dare you! Do you have any idea how embarrassed I was?"

My expression turned deadly, and he must have been so shocked to see my face twist into something besides placation, that he really did sit down and shut up.

"Surprised to see me, Dad?" I asked, fully aware he hadn't said anything yet. "Or are you and Jake having a plotting session, trying to figure out how best to deal with the latest Caroline complication?"

My father scowled. "Don't be so dramatic, Caroline."

"You don't look very happy to see me."

He frowned. "You can't just run off like that, leaving hundreds of your wedding guests waiting, hiding for nearly a year, then expect me to be happy when you finally decided to come home."

"Then why offer a reward?" I asked.

He waved his hand in dismissal. "It's the thing people expect."

"Exactly," I said. "It's the thing people expect. Which is why it raised eyebrows when you waited so long to offer it."

"Where have you been, Caroline?" he asked in disgust.

"You're not going to even pretend to be happy to see me?" I asked, feigning surprise.

"Where have you *been*?"

"I think you know," I said. "Maybe not *exactly* where this last time, but you knew where I lived from August until October."

His face hardened. "I don't have time for games."

"Of course," I said, waving toward him. "I can see you're busy, smoking your cigar and plotting your next move with Jake. Although you strike me as the mastermind of the whole operation, while Jake just does as he's told. At least that's the way I understood it the night of the wedding rehearsal. When you and Jake were *plotting* then."

Jake's face paled and he turned to my father. When he got no cues, he must have decided to go on the offensive. "What the hell happened to you, Caroline?" Jake asked. "You used to be so..."

"Docile?" I supplied. "I realized that if I wanted to survive, I needed to grow up and stop being a doormat for you and my father to wipe your feet on."

"If you were unhappy, then you should have just said so, Caroline," Jake said, rolling his eyes, but he looked like he was trying to play the role my father had assigned to him. The sweat dotting his forehead told me that he was freaking out. "We could have gone to couples counseling."

"I'm not sure a therapist would have been able to help us work through the fact that you and my father were plotting my murder."

"What the hell are you talking about?" he demanded, but he was too outraged to pull it off.

"Look," I said. "I'm done playing games. I'm tired of hiding, so let's cut the bullshit, okay? Why pretend it didn't happen, Jake? After so many years of playing me, the least you can do is own up to your actions."

"Own up to what?" he asked, trying to play stupid, but it sounded just as forced as everything else he'd said. Guilt was written all over his face.

He was obviously the weak link here, no surprise, but I needed to work on my father too. Getting him to confess was my real mission.

"But you've learned from my dad, and he likes to slide under the radar. He's pretty good at it. Somehow, he escaped arrest when the Hardshaw Group went down."

Jake's eyes bugged out, but my father sat still, watching me like I was a panther on the prowl. The thought gave me more satisfaction than it probably should.

"But you've made getting away with things an art form," I continued. "First Grandma and Grandpa's accident and then Mom's."

"What the devil are you talkin' about?" my father asked, his Texas drawl coming through. His drawl had always become more pronounced when he was stressed.

I was pleased about that too. Maybe I *could* pull this off.

"I heard you," I said evenly. "The night of the rehearsal dinner, I heard your plans for me. I also heard you admit to what you'd done in the past, kind of like a résumé to convince Jake that you could both get away with killing me. I was terrified, so I ran. And you must have known that was why I ran. I'd heard something, but how much? Had I heard that you intended to bring Jake into Hardshaw after our marriage?" I made a face. "I did. Had I heard you admit to murdering my mother? I heard that too. But you didn't know. I was a risk, and you'd planned on killing me after Jake was firmly situated both on the Blakely Oil board and the Hardshaw Group, so you might as well move up the timetable and kill me sooner. Only I wasn't so easy to find."

My father's jaw hardened, his eyes glittering with anger.

"You've lied and manipulated me for most of my life, so let's just cut the bullshit. Do me the courtesy of telling me the truth. Just once."

Neither one of them said anything.

Crap, I had a feeling I was running out of time. I'd made a lot of accusations, and while Jake looked suspicious as hell, my father just looked pissed.

"I know you questioned Aunt Tiffany after I disappeared," I said to my father. "And I'm sure you know I met with Uncle William yesterday morning in Atlanta. He's terrified of you, so good job there."

A slight smile tipped up the corners of his mouth. "I can't help it if your uncle has always been a weak man."

Finally. A response.

"He must not have been too weak if he dared to have an affair with your wife," I said.

My father's nostrils flared.

"Yep. Heard that part too. You know, I've heard quite a few fun things when you thought no one was listening. I also learned that you were a partner in Hardshaw, and you were planning to expand to Henryetta, Arkansas. You figured out that I was hiding there, so after the big bust there, I ran again."

"Why in the world would you go to some Podunk town in Arkansas?" Jake asked.

"Trying to find answers. Trying to find out what was so damn important about you joining Hardshaw that made it worth murdering me."

My father gave me a sympathetic look. "Now, Caroline. I realize that you're unwell, and that this has been an ordeal." He got up from his seat. "We'll just call Dr. Marlow, and he'll give you a sedative to calm you down."

Shit. I should have realized it wouldn't be easy to get him to confess. I'd just reminded him of a lesson he should have learned years ago: you just never knew who was listening. He wasn't making that mistake again.

Jake got to his feet too, and I realized I needed to put the screws on him. Maybe Wyatt had been right about all of that.

"I know about the insider trading charges," I said. "They'll indict you soon."

He stopped in his tracks, his face frozen in shock. "What the hell are you talking about?"

"You can work out a deal if you tell them what you know, but they'll want to know it all. You might be able to get all the charges dropped, though, because I'm sure you know *a lot*."

His eyes bugged out, and he turned to face my father. "I have no fucking idea what she's talking about, Randall. I swear to God."

My father tilted his head, his gaze narrowing, then pulled out his phone and pressed a number. "Ricketts, I need you to come to my study. Now." He hung up the phone, and Jake took a step backward.

"I swear to God, Randall," he whined. "I haven't done any insider trading. I'm not that stupid."

My father ignored him and turned to me, reaching for my arm. "Caroline, let's go to your room and rest until Dr. Marlow gets here."

I backed away from his reach. "Keep your hands off me!"

"Now, Caroline, we've all been worried about you. Let your father take care of you."

"For God's sake," I said in frustration. "For once in your life, would you tell me the truth about *something*?"

He stopped and studied me for a moment, giving me a patronizing smile. "Okay." He held out his hands in surrender. "I admit, I'm not your biological father, but I didn't find out until after your mother died, and as far as I'm concerned, you're my daughter, Caroline. The DNA didn't make a difference."

"I *heard* you when I was a kid," I insisted. "When you and Mom were trying to have another baby, and you found out you were sterile. You realized Uncle William was my father, and you were *furious*," I said, fighting to keep my voice cold and calculating, because God knew if I showed any sign of emotion, I'd be labeled a hysterical woman, but Jake over there, who was currently pacing as if his life depended on it—and it probably did—would just come across as stressed.

Fuck that.

"Then, *coincidentally*," I continued, "my mother died soon after you found out, and you pretended like I was dead too."

His face softened with understanding and sympathy. "I realize that losing your mother was difficult for you. And I was so lost in my grief, I'm sure I wasn't there for you like I should have been."

"I was eight years old and motherless, and you fired my nanny."

He started to say something but struggled to come up with a response. I mean, how did you come back from that?

"I would never sell you out, Randall. You *have* to believe me," Jake pleaded.

My father gave him a warm smile. "Of course not, Jake. You're like a son to me."

Jake's face fell, and he bolted out of the room.

Obviously, he'd seen how my father treated his children.

"So it's just you and me, Dad," I said. "Why don't we cut the crap and be honest? We both know I probably won't see the sunrise tomorrow morning, so why not just confirm what I already know? You killed my mother."

He grinned. "You know it's easy—"

A scream sounded from somewhere downstairs, echoing off the marble floor, quickly followed by multiple gunshots. The screaming stopped. One more shot exploded.

I jolted in shock.

The sound of heavy footsteps running down the hall ended at the door behind me.

"She's airing this live!" a man shouted, then hands grabbed my arms in a vise grip. A man dressed in all black moved in front of me and started feeling me up, finding my phone in my bra. He pulled it out, then threw it to the ground and shot it with his semi-automatic rifle.

My father's mouth had dropped open in shock, but then he turned to me with a look of pride, something I'd been searching for my entire life. Now, it disgusted me.

"You shot a hole through the rug," I said in disgust. "You couldn't just turn it off?"

"Ricketts is a man of action," my father said.

And apparently not enough brains to realize that pressing an off button might be more sensible than blowing a hole through a twenty thousand-dollar wool rug, and probably the walnut floor below.

"You almost got me, Caroline," my father said, still looking proud. "Maybe I underestimated you."

"Too bad it took you thirty-two years to figure that out."

"You know," he said, "if you hadn't made that whole exchange public, we might have been able to work something out."

"Liar," I said in disgust. "I'm not your flesh and blood, so I'd never be good enough for you."

He made a face. "You're right. Your mother was a lazy whore who spread her legs for anything with a swinging dick, and the apple doesn't fall far from the tree."

"Some goddamned father you are," I heard a deeply familiar voice say behind me. "Because you don't know my girl at all."

I glanced over my shoulder, my mouth falling open when I saw Hank standing in the doorway, holding a semi-automatic weapon on my father.

What was *Hank* doing here? How had he found me?

And where in the world had he gotten a semi-automatic weapon?

The two security guards swung their guns on him.

"You take one shot at me, and I'll blow your boss to smithereens," Hank said. "I suspect you'll be missin' out on your year-end bonus if I do."

My father looked Hank over, taking in his crutch and empty pant leg, and laughed derisively. "Who the hell are you?"

"A better damn father to her than you'll ever be," Hank grunted.

A lump filled my throat. I loved this man.

"What the hell do you think you're gonna do with one leg?" my father asked with a laugh.

"Last time I checked, I don't need two legs to pull a fuckin' trigger."

I *really* loved that man.

"You're gonna let her go," Hank said, "but first, you're gonna tell her what she wants to know."

"I don't think so, you redneck," my father said with a laugh. "Ricketts, take care of—"

The guard who'd pulled the phone off me, lifted his gun, and a succession of loud shots instantly went off. I flinched, terrified for Hank, but Ricketts hit to the ground with a thud.

The other guard lifted his gun, and Hank shot him next.

"Okay," Hank grunted as the second guard fell. "Now that we've established that I do *not* need two legs to pull a trigger, let alone make my shot, I hope I have your full attention."

My father held up both hands in surrender. "What do you want to know?"

I turned to face him, still stunned by what had just happened with the guards. If age had slowed Hank's reflexes, he must have been like greased lightning in his youth. Then again, I'd seen him kill a man with the same skill and lack of hesitation within my first few days in Drum. He'd done that to protect me too.

From the look on my father's face, he was just as impressed. Or at least appropriately intimidated.

I turned to him, my anger igniting. "I want you to admit that you killed my mother."

He quirked his mouth in disgust. "You've come all this way, from the timid mouse to—" he gestured to me, "—the woman you are in front of me, and you're asking questions you already have the answers to. Why aren't you asking for information you don't have?"

He was right. He'd already admitted that he'd killed her. He just hadn't admitted it to *me*. Before, I'd wanted it recorded so he could pay for his sins. But my phone had been destroyed, and this conversation was just between him and me.

I looked him in the eye. "Where's the Rucker locker?"

He blinked and gave me a hard stare. "I don't know what you're talking about."

"That's a bald-faced lie," Hank said. "Try again. Where's the Rucker locker?"

"Where'd you hear about that?" my father demanded.

"It doesn't matter," Hank said. "It's not gonna do you much good in prison, so you might as well tell your heir."

My father's face paled. I wasn't sure what part of that statement scared him—Hank's intimation that he was going to prison, or the possibility that I would inherit all his possessions.

Oh shit. *Would* I inherit his fortune? He'd probably willed it all to

Jake, but based on the screams and gunshots below, Jake might not be in line to get anything anymore.

But my father wasn't cooperating, and I didn't feel like going through the whole spiel of threatening to shoot him if he didn't. I had another trick up my sleeve.

"I know about the Carpathian deal."

His eyes bugged out as he turned to me. "How do you know about that?"

"I just do, so unless you want the people from the Carpathian deal involved, you'll tell me where the Rucker locker is."

His eyes narrowed. "I don't believe you. You'd never be involved with anyone as ruthless as them."

Hank barked out a snort. "How the fuck can you say that when she's with me? You have no idea what this woman is capable of because you completely wrote her off. Now where the fuck is the locker?"

My father eyed Hank, then looked me up and down. "I need to go to my desk. The information is on my computer."

An antique walnut writing desk sat in the corner of the room, with an Apple desktop computer arranged on one side. Hank motioned with the gun for him to walk over to it.

My father sat behind the desk and booted up his computer. I walked to the side of the desk to keep an eye on him, while Hank moved deeper into the room, balancing on his crutch.

My father started typing on his screen, and I watched to make sure he was opening a file and not sending a text or making a call. To my surprise, he opened a file with very little text and then hit print. The printer behind him whirred to life. He spun in his chair and snatched up the paper, handing it to me.

I took it and looked it over. None of it made sense. I didn't see any directions; instead it looked like computer code.

"Happy?" my father asked.

Was I? Skeeter hadn't given me any information about what I needed, so I could only presume this was it.

"You're going to want an envelope for that," my father said, then reached for his desk drawer. "Sensitive information like that should be protected."

I lifted my gaze from the paper to watch him. He reached into the top drawer on the right side, and before I could react, my father pulled out a handgun and fired.

CHAPTER THIRTY-THREE

MARCO

My phone may have told me I was fifteen minutes away, but the Dallas traffic had other ideas. It had slowed to a crawl, cars stacking up around me. I considered abandoning Hank's vehicle right there in the middle lane and running on foot, but I was still seven miles away. Still, it might have been faster than this.

I found Jed's number and called him, praying he was already on his way to Carly.

"Please tell me you're almost to Blakely's house," I said as soon as he answered.

"No. The runway at the airport was closed, so we were diverted to another airfield. We're about to land now, but we still need to get vehicles. We'll be at least another twenty minutes, but I'm guessing closer to a half-hour."

Fuck. Carly had already been at Blakely's house for at least five minutes. She could be dead by now.

"I'm stuck in traffic, but I'm about to push my way to a shoulder and drive to an exit."

"Do it," he said. "We need to get to her. We have to stop Wyatt."

"Hank's on his way too. He may actually beat me there."

"Good. Let me know what you find when you get there," Jed said. "Marco, if you love her..."

"I won't let anything happen to her. I swear it."

I hung up and laid on my horn, forcing my way across three lanes of traffic until I was on the shoulder, then sped toward the exit two miles ahead. After I got off the highway, I let my phone lead the way.

Ten minutes later, I drove past the front of the house, planning to do a drive-by to find the best entrance, and saw the front gate was partially open. I'd figured going in through the back would allow me a stealthier entry, but I was short on time. Going through the gate could save me precious minutes...or it could be a trap. I was willing to take the chance.

I parked at the curb, ignoring the no parking signs posted along the front of the wrought iron fencing marking the front property line. Grabbing my handgun out of the glove compartment, I got out and rushed through the gate—which appeared to have Hank's handiwork all over it from the pry marks on the latch—and up the hundred-foot-long drive to the two-story brick home that had just about every window illuminated.

The front door was partially cracked open, and an ominous feeling settled over me. Both entry points were open? Maybe it really was a trap. Or maybe Hank had purposefully left the doors open to make it easier for me to get in.

Holding my weapon like I would on any other bust, I entered through the front door.

The smell of blood hit me as soon as I walked inside. Two men were down. A man dressed in all black tactical gear was slumped against a wall next to a hallway with a bullet hole dead center in his forehead. He'd left a bloody streak down the wall as he collapsed. The other man was dressed in a polo shirt and khakis. He was lying face-down on the marble floor in a large pool of blood. Based on the bloody spots on the back of his polo shirt, someone had shot him multiple times. I figured he was a civilian causality, but then I recognized him.

Jake.

Jesus Christ, had Hank shot his way in here and taken out Jake? How was I going to get him out of this?

Even though I knew it was unlikely he was alive, I squatted next to him and pressed two fingers to his carotid. He was still warm but didn't have a pulse.

Shit. This was already messy. I hoped to keep it from getting any messier.

I got to my feet, trying to figure out where to go next, but then I saw a man in a suit hanging back in the shadows in a room off the entryway. His eyes were huge, as though he was terrified he'd be the next victim.

"I'm looking for Carly," I said in a firm tone. "Where is she?"

"Are you another friend?" he asked, barely above a whisper.

Another friend? Hank must have told him he was looking for Carly. "Yes. I'm her boyfriend. Where is she?"

Relief filled his eyes. "He was going to kill her, and then I heard gunshots from up there. I'm frightened he—"

My heart slammed into my ribcage. Gunshots? "*Where?*" I demanded.

"Upstairs," he said, pointing upward with his index finger. "In Mr. Blakely's study. Down the right corridor after you get to the top of the stairs."

"How many gunmen?" I asked.

"Three, including Carly's other friend, but there are probably more on the way."

I bolted for the stairs, past the fallen guard. The man didn't have any visible weapons, and something told me Hank had liberated them. When I reached the top, I stopped and glanced down both corridors, then began to slowly make my way down the right one, my gun at the ready. I considered clearing rooms as per procedure, but I heard voices further down the hall and light spilled out into the dimly lit hall.

The room was eerily silent as I approached, but when I'd almost reached it, two gunshots exploded, and I heard Carly scream.

My training kicked in, making me ignore the impulse to just burst into the room to protect her. Bursting in could get me killed, and if I were dead, I wouldn't be any good to her.

I stood with my back to the open door, my gun upraised, and my arms held tightly against my body. "Carly?" I called out, grateful my voice didn't break. "Are you okay?"

"Marco?" she called out in surprise. "I'm okay. You can come in."

My knees almost gave out, but I turned the corner, still on alert as I took in the scene in front of me.

Two additional guards lay dead on the floor. Hank was in the middle of the room, holding a semi-automatic rifle. A man was slumped over the desk, his hand holding a gun, a puddle of blood forming underneath him. Carly stood to the side of the desk, holding a sheet of paper in her hand, looking startled. She stared at me with wide eyes, then cried out, "Hank!"

As if on cue, Hank fell to the floor.

Carly threw the paper onto the desk and rushed over to him, and I saw a red stain blossoming on his upper shoulder.

I needed to make sure the threat was neutralized, then triage the patient. I went for the man at the desk, feeling for a carotid pulse, but his open, vacant eyes made it clear he was dead, the cause of death a bullet hole in the center of his forehead.

That and a good portion of the back of his head was missing.

Randall Blakely. Thank God he wouldn't be threatening Carly anymore.

Carly was on her knees at Hank's feet, starting to cry as she put her hands over his wound to stop the bleeding. I grabbed a velvet throw pillow and hurried to Hank's side. He looked pale and shaky.

"Are there any more of the bastards?" Hank asked in a raspy voice.

"No," I said, then glanced over at Carly holding out the pillow. "Let's use this to apply pressure to his wound. We need to call 911."

She took the pillow and pressed it down hard on Hank's shoulder.

"Ow!" he shouted. "Be a little more gentle! And I don't need no damn hospital. You can't be callin' 911."

"He's right," a man said as he entered the room, filling the doorway. He was dressed in tactile gear too and looked like he could do a bicep curl with a small car, but I quickly dismissed him as a threat when I recognized his voice.

"Happy to see you, Jed," I said with a grimace, "but we're not covering this up."

"Agreed. It's too big to cover up, but we can minimize the fallout." He glanced at Carly. "Did you accomplish whatever you needed to do for Skeeter?"

She nodded, glancing up at him. "The paper on the desk."

Jed walked over and started to pick it up, but she got to her feet and snatched it from him. "That's not for you."

I took over applying pressure to Hank's shoulder.

Reaching into the partially open drawer, Carly pulled out an envelope, hastily folded the paper, smearing it with blood, then stuffed it into the envelope and sealed it. She pressed her thumb over the middle of the flap, leaving her bloody fingerprint like a wax seal, and slipped it into her jacket pocket, giving him a defiant look. "It's Skeeter's. I told him not to underestimate me."

He nodded sharply. I was worried he'd be pissed, but he gazed at her with respect. "Anyone who does is a fool."

"Where's Wyatt?" I asked, glancing around the room.

"At a gas station somewhere off I-30," she said, nonchalantly.

"He didn't come with you?" I asked in disbelief. "He didn't turn you in?"

"No," I said. "In fact, he told me I didn't have to do this. He offered to drop me off at the airport so I could fly home to you."

I stared at her in disbelief.

I didn't have time for more of a reaction, because four more men appeared in the doorway.

"The property seems to be cleared," one of them said. "What now, Jed?"

Jed was leaning over the very dead Randall Blakely, tapping on the computer without any sign of distress. "Make sure there's nothing implicating Carly or Hank. If so, we clean it. Once we're certain, we call it in."

"You don't think the staff has already called it in?" I asked.

"I know for a fact they haven't," Jed said. "They're used to keeping their heads down and letting their boss deal with violence." He looked up at me with a half-grin. "And I asked the butler."

"Is Reginald okay?" Carly asked.

"He looked fine," Jed and I said at the same time.

"We can't wait to call this in," I protested. "He's a public figure. They'll be going over the scene with a fine-tooth comb."

"We only need a few more seconds," Jed murmured, his gaze on the screen while he tapped on the keyboard. He studied the screen for about ten more seconds, then glanced up at Carly. "Carly, call 911."

"What do I tell them?"

"The truth up until the moment me and my men showed up. I just erased us from any security footage, and what it captures of the rest of you will prove your actions were justified." He stood upright and came around the desk, giving her an affectionate smile as he stopped in front of her. "I'm proud of you. The live Facebook feed was brilliant."

"You knew?" she asked in surprise.

"What Facebook feed?" I asked.

He grinned. "Neely Kate saw it. She's been monitoring your social media since we first found out who you are. In fact, I'm surprised someone hasn't notified the authorities yet, even more reason for us to run." He pulled her into a quick hug. "You better come see us when this is all wrapped up." His gaze swept around the room. "You're gonna be stuck here in Dallas for a while."

"You think we'll be in trouble?" she asked, looking scared.

"With the law? Nah. NK says it was obvious Blakely was hostile. Hank may have some explaining to do, but he was obviously acting in self-defense. The footage and the butler will back that up. And while the legal mess may take a few days, the issues with the company will likely take a lot longer. You're about to have a PR nightmare on your hands."

Carly's wild eyes swung to me. "I don't want the company."

"You may not want it, but I suspect you'll get it," he said. "You're going to have to act fast to keep it from sinking."

"I don't know how to do that," she said. "I can't do any of that."

"We'll deal with it, Care," I said, taking her hand. "We'll deal with all of it together."

CHAPTER THIRTY-FOUR

CARLY

Two weeks later

I looked up from my desk at Blakely Oil, bleary-eyed from too much paperwork. I'd appropriated Jake's junior executive office, but I'd gotten all new furniture to replace his things.

"Take a break, Care," Marco said from across the room. "Let's get some fresh air.'

"It's nearly a hundred degrees out there," I said with a laugh, pointing to the glass windows of the forty-first-floor office. "I don't want to be all sweaty when we meet Max for lunch."

Max was coming to see us for the first time since I'd been kidnapped by his brother. He'd been dealing with the fallout of his father's arrest and his mother's disappearance. Emily still hadn't been found, but so far no one had tried to blackmail Max about the missing files.

Bingham hadn't breathed a word about his brother's body being found, but Hank suspected Bart Drummond would soon meet his demise while sitting in the county jail. No one would lose much sleep over that, including Max, who insisted his father would only get what was coming to him.

No one had heard from Wyatt since I'd last seen him in my rearview mirror.

Marco would be leaving soon to pick up his best friend from the airport, and I knew he was worried about leaving me alone for the first time since any of this started.

"Then let's go to the coffee shop downstairs," he said. "A change of scenery will do you good."

"Why would we go downstairs to the first floor when there's a full coffee bar down the hall?" Truthfully, the staffers looked horrified every time I used it, like they thought they were all about to lose their jobs because no one had read my mind and made coffee for me.

These people had a lot of PTSD to unpack. I understood that all too well.

"Carly, you've done nothing but pore over paperwork for days. The company's not going anywhere. You can take a ten-minute break."

"I know," I said with a frown. "It's just that so many people are depending on me..."

The night my father had died, I'd been more than willing to let his company die with him, but then I realized how many people Blakely Oil employed. To let the company die would mean taking away the livelihood of thousands of people. So after we'd all been cleared legally, my father's attorneys had sat me down and told me that Jake had become the beneficiary of my father's estate. Both of them were gone, however, so it had gone to me.

The feds were still investigating to see if the company had been part of my father's illegal dealings, but so far it looked clean.

Which meant I was now a multi-millionaire.

Only I didn't want any of it.

I'd feared Marco would go running for the Tennessee hills, but he'd vowed to stay with me until I sorted out the company and decided what to do.

"What if I decide I want to run it myself?" I'd asked a couple of nights ago while we ate a late dinner in our rented corporate apartment.

"I'll start looking for a job in Dallas," he said, then shrugged. "Or Fort Worth. Maybe I could get a job with Duncan Morris's security firm."

"You don't want to provide *my* security?" I asked, feeling hurt.

"Of course, but I don't know if I can work for your company, Carly. I'd rather earn the job, not get it because of who I'm engaged to."

I'd stared at him in confusion. "Engaged?"

Looking nervous, he'd pulled out a diamond solitaire ring from his pants pocket. "It's not big. Nothing like what you can buy your—"

I shut him up by leaning closer, grabbing his shirt, and kissing him. "I love it," I said when I pulled away, my eyes stinging with tears. "It's perfect."

"I know the timing sucks, but I want you to know I'm with you, no matter what you decide. I don't care where we are, as long as we're together."

I was wearing the ring now, smiling to myself as I glanced down at it. I felt lucky to have such a generous man, but I had to make sure he was happy too.

I didn't think I wanted to run Blakely Oil indefinitely. Only until I made sure it was on solid footing and I found someone trustworthy to run it.

I pushed my chair back and got up, slowly walking toward Marco, who was sitting on the leather sofa, papers spread out around him. "You know," I said seductively. "I could take a ten-minute break."

His gaze traveled from my three-inch pumps, up my bare legs, then my skirt, my silk blouse that clung to my breasts, then up to my face. "Why, whatever do you have in mind, Ms. Moore, and do I need to sign paperwork in HR first?"

I burst out laughing, then sat sideways on his lap. "Don't ever leave me, okay?" I whispered.

His gaze grew fierce. "I love you too much to live without you. We're gonna grow old together, Carly Moore."

I gave him a soft smile, then played with the collar of his dress shirt for a second. "I've been thinking about what I'd like to do for the rest of my life, but you're a part of the rest of my life too," I said quietly. "What *you* want matters. In a perfect world where money is no concern, what would *you* do?"

He smiled, tucking a strand of hair behind my ear. "That's easy. Be with you."

"No," I reprimanded. "You prefer the country life to the city, right?"

"Carly," he groaned. "I'll do whatever—"

"Marco. Country or city?"

"Country," he said with a sheepish look.

I pressed my forehead to his. "Me too," I whispered.

He kissed me. I let my lips linger on his, feeling grounded again, before I pulled back a few inches to look into his eyes.

"That gunshot has taken a toll on Hank. I'm worried about him."

"Me too," he admitted.

Hank was still in Dallas, but he wasn't happy about it. He was part of the reason Max was coming to visit. He planned to take Hank back to Drum with him when he left.

"I've spent a lot of time without a dad," I said, my eyes stinging with tears. "I've been looking for Hank my entire life. I want to enjoy every minute he has left."

He chuckled. "You know he's too stubborn to die on us and probably has another twenty years left, right?"

"Good," I said, playing with the button on his shirt. "So what do you think about moving back to Drum?"

His eyes widened in surprise.

"Feel free to tell me no," I said. "This is just a suggestion, but I was thinking maybe we could use some of my money to buy the land around your house. That way we can make it a little more secure and private."

He smiled. "I think I could live with that."

"There's more."

"I'm listening."

I bit my bottom lip. "What if we open a charter school? But it's open to *all* kids, and we make sure they get the education they need to get out of Drum if they want, but I was thinking… Maybe we can help make it a place they'd want to *stay*. I mean, between Max continuing with his father's resort and the increase in tourism, we can use some of my money to give grants and small business loans to the people in Drum so they can secure their own futures." I gave him a hopeful look. "What do you think?"

He watched me for a long moment, his eyes gleaming, then a slow smile spread across his face. "I think I'm the luckiest damn man alive."

I laughed. "Yeah, I think so too."

He swatted my bottom and then kissed me. I reveled in our happiness. Maybe, just maybe, everything would be great after all the shit we'd been through.

Pulling away reluctantly, he said, "I need to leave to pick up Max. You still want to meet us at the restaurant?"

"I think that works best," I said. "That way I can get more work done."

He nodded, then held onto me as he stood. He set me on my feet, giving me one last lingering kiss. "I love you, Carly."

I reached up and caressed his cheek. "I love you too, Marco." I dropped my hand and took a couple of steps back. "See you in an hour."

He walked out the door, and it instantly felt like a hole had opened in my heart, which was stupid. I'd see him in an hour.

Besides, I had more than enough work to keep me busy until it was time to go.

I sat back down at my desk and got back to work. I was lost in my reading when my office door opened, and my assistant walked in. Janie had been Jake's assistant, but she hadn't had a problem with me slipping into his place. If anything, she'd seemed grateful.

"Carly, I just wanted to remind you of two things," she said as she approached my desk. "One, I'm taking a longer lunch today for my doctor's appointment. And two, you have your own lunch to attend, and I won't be here to remind you to leave." An apologetic look covered her face. "Mr. Roland suggested I remind you to set an alarm on your phone."

I laughed. "Thank you for both reminders, and Marco's right. I need the alarm. I hope your appointment goes well."

She waved her hand. "It's just a checkup, but I know I'll be sitting in the waiting room for a while before they call me back. But thank you."

"Of course."

She headed out the door, then stopped and turned back. "I know you're not sure if you actually want to continue working here, but you're really good at it, and everyone has your back." She hesitated. "Your father wasn't an easy man to work for. Your leadership could help this company thrive."

"Thanks," I said. "I'm still trying to decide, but you have no idea how much that means to me."

My father had thought I couldn't handle running his company.

Look at me now.

I set the alarm on my phone, then got back to work. About ten minutes later, the door opened again, but I was concentrating on a paragraph about clean energy proposals my father had shot down. I didn't want to lose my place, so I kept my gaze on the paper and said, "Are you back already, Jane?"

"Sorry," a male voice said. "Not Jane."

My gaze jerked up, and I was shocked to see the man standing in front of me. My nerves stood on end. Why was he here and what did he want? There were multiple warrants out for his arrest.

"Wyatt! Where have you been?" I asked. "Max has been worried sick."

"I'm sure he has." He took a seat in the chair in front of my desk. He was wearing jeans and a light gray T-shirt, a Texas Rangers ballcap on his head. He took it off and tossed it in the seat next to him. He looked like he hadn't shaved since I'd last seen him, and his hair was a mess from the hat. Dark circles underscored his eyes. He looked like a mess, but then again, he was a wanted man.

Max had tried multiple times over the past two weeks to call him or email him and it was like he'd fallen off the face of the earth.

I couldn't believe he was here with me now.

His upper lip curled, and he gestured toward me. "Look at you. You're all corporate now."

I made a face, trying to hide my nervousness. "I'm not sure for how long. I'm about to start the search for someone to run the company in my place."

He leaned back in his seat, resting his hands on the chair arms. He looked exhausted. "Got everything you wanted, huh? I saw in the news that they reopened your mother's accident and ruled that her car had been tampered with. And a mechanic came forward and said his boss, long since dead, had been paid to tamper with her brakes."

"Yeah."

"And you inherited your father's estate."

"I never asked for it, but yeah."

He nodded, barely concealed anger burning in his eyes.

I knew where this was headed. That's what Wyatt had wanted from his own father, and he hadn't gotten any of it. Now he was on the run, and Max was running everything instead.

I drew a shaky breath. "Wyatt, I'm sorry about Jake. I had no idea he'd be at my father's house when I showed up. I truly believed you could get what you needed from him and set things right for your

FBI deal. But they're uncovering new things every day. I'm sure we can find something else for you to use."

"There was no FBI deal, Carly," he shouted, his expression darkening. "I just made all that shit up to get you to go along with me! Have you really not figured that *out*?"

While it had occurred to me many times, I hadn't let myself believe it. He'd been so nice to me at the end. I'd wanted to believe he was redeemable, if only for Max's sake. The next words escaped me in a whisper. "If you lied about that part, then did you lie about Ruth?"

The Hensen County Sheriff's Department still hadn't solved Ruth's murder, and I hadn't told them that he'd confessed to me and then recanted. Neither had anyone else, although Hank was itching to find Wyatt and deal with him himself.

"Yes," Wyatt said through gritted teeth. "I lied about that part too."

I stared at him for a long second. "So you really did kill Ruth?" I asked in disbelief.

He held my gaze, a hard gleam filling his eyes. "Yes, I shot her and Franklin too. And it wasn't a fucking accident."

I felt like I'd been punched in the solar plexus. He'd murdered Ruth, then pretended to be upset about her death and implicated Franklin. He'd done it so convincingly, with such believable grief.

"You're proud of it?" I asked, my anger building.

"Of course not!" he exclaimed. "But I had a job to do, and she got in the way."

"You killed your friend for absolutely nothing!" I shouted. "You weren't working for the FBI. You were working for *yourself*?"

"Working for myself is *not* nothing!" he shouted. "I've gone my whole life without anyone looking out for my best interests! It was time to take matters into my own hands."

"Are you *kidding* me?" I shouted, getting to my feet. "You had your whole damn life handed to you on a silver fucking platter!"

"As long as I followed my daddy's rules," he said bitterly. "You had a taste of that with your own father, Caroline. How did *you* like it?"

"I didn't," I sneered. "But I sure didn't lick his bootstraps for years, hoping he would finally give me his stupid approval."

"And yet you still got the prize," Wyatt said bitterly, gesturing toward me. "Look at you now."

We stared at each other for several long seconds—me still standing behind my desk, him sitting in the chair. Finally, I said, "What do you want, Wyatt?"

"I want what's mine."

I lifted my hands in frustration. "And how exactly am I supposed to make that happen?"

"That's just it," he spat out. "You can't. But if I can't have everything I want, then why should you?"

He sat up in his chair and pulled a gun out from behind his back, pointing it at me.

My heart sank. I had a feeling there was no talking myself out of this, and Max and Marco were on their way to the restaurant.

"You may have gotten past security," I said, trying to keep my voice even, "but you're still on cameras. Everyone's going to know that you did this. You're not going to get away with it."

"Maybe I know that already," he said with a shrug, then sat back down. "Maybe I don't walk away from this. Maybe I just destroy what I can on my way out."

"Like a kid who can't have his way," I said.

"If you like."

I sat down in my seat, my brain scrambling to catch up. There had to be a way out of this. I just needed to figure out what it was.

"So Jake," I said slowly. "There was no insider trading deal, was there?"

He laughed. "No."

"But you knew specifics," I said in confusion.

A smug grin spread across his face. "I didn't. I provided a few things, and you filled in the gaps. I suggested it was a friend from high school who was in finance, and you supplied the name. I just threw out some random pharmaceutical company. Seemed like a guy like Jake might do such a thing. And I knew he had investments. Guys like him always do."

I remembered the surprise and horror on Jake's face when I'd mentioned it, the way he pleaded with his father to believe that he hadn't done it.

Oh my God.

Had *I* gotten Jake killed?

My head swam, and it took a few seconds for the ringing in my ears to resolve.

I stared in dismay at the man in front of me. His eyes burned with anger and absolutely no remorse.

They were also filled with determination.

Wyatt had no intention of letting me leave this room alive.

"So how's this gonna work?" I asked, sitting back in my seat. "Murder/suicide? You're going shoot me in my chair and then turn the gun on yourself?"

"You took everything away from me, Carly. Don't you see that? You have to pay."

"Or..." I said slowly, my anger rising. "You could actually take responsibility for your own actions for once in your goddamned life."

"You really don't see how you fucked me over?" Wyatt asked, outraged.

I shook my head. "Yeah, Wyatt. I fucked you over," I said wearily.

He lifted his chin, looking smug. "Thank you for admitting it."

My mind went over those last hours in the car with him. How sweet he'd been. How solicitous. It had all been a lie?

"You planned to turn me in to my father all along, didn't you?"

"Yep. Once my anger died down, I realized that you were always a sucker for the underdog. If I could convince you that *I* was the underdog, I could get you to work with me to go to Dallas and confront your father."

"Did you ever plan to go to Jake's?"

He chuckled. "Hell, no. When I told you I planned to drive you to the airport, I was planning to drive you to your father's house."

None of this made sense. "So why did you say you'd do it without me? Why not just give me a pep talk?"

He leaned forward, his face hard. "Because over the last eight months I learned that you *really* hate being told what to do." A sparkle filled his eyes. "Want to get Carly riled up? Tell her she's incapable of doing something. Another byproduct of us poor rich kids trying to make our daddies love us," he said mockingly.

I stared at him in disbelief.

"So when I suggested you just give up, I knew you'd be inspired to confront him. The one thing I hadn't counted on was you leaving me behind." He laughed bitterly. "I suspect you did that to protect me, huh?"

"Yeah." I supposed a small part of me did do it to protect him, but I'd mostly done it for myself. Still, I didn't see any point in deflating his massive ego. "We've already presented enough evidence to prove my idiocy, so..."

"Anything else you want cleared up?" he asked. "It's a shame no one else will know my true genius."

"Jerry?"

For the first time, he showed a modicum of remorse. "Collateral damage."

I nodded. There was no doubt Wyatt was a monster. But I still had trouble believing he'd purposely killed him.

"Heather?" I asked.

"You know I didn't kill her," he said with a short laugh. "You found out the truth on that one."

"What about Rodney Bingham?" I asked. "Did you know your father killed him and buried him practically in your backyard?"

"No, but it doesn't surprise me. It also doesn't surprise me that he was Rodney's father." He held his hands out. "Anything else?"

"Did you steal the file from Ruth?"

"Of course," he said, "but I burned it. I may be an asshole, but I'll protect my brother."

"You couldn't tell him that and give him some peace of mind?"

"Nah. It'll do him good to sweat a little."

So much for protecting his brother.

We were both silent for a moment.

"So this is it, huh?" I asked. "I break down in some random town, and you're the one to pick me up and start me on my new life. Then, less than a year later, we're sitting in some high-rise office in Dallas, and you kill me and then yourself?" I made a face. "Sounds like some kind of Lifetime movie. I never thought you were so cliché, Wyatt."

"Shut the fuck up, Carly," he growled.

"Why? You're going to shoot me anyway, so I might as well say whatever the hell I want."

The alarm went off on my phone. I reached over and turned it off, my gut sinking. That explained why no one had come in to check on me despite all the shouting. Everyone was at lunch.

No one was coming to save me. I'd have to do that myself.

"What was that?" Wyatt asked.

The key was to get him good and pissed so he'd lose control, then push his buttons even more.

I looked at him like he was a two-year-old. "I realize things are backward in Drum, but this is what is known as an iPhone." I opened my top drawer and tucked my cell phone inside, then blindly grabbed my letter opener and kept it hidden as I closed the drawer.

"You think I'm playing?" he asked, getting to his feet.

"No," I said, giving him a hard stare, drawing his gaze up to my face as I situated the letter opener in my hand. "I know you're going to kill me, but you're not going to kill yourself. You're too big of a coward to do that."

"Shut up, Carly."

I stood and started to walk around the side of the desk. "I may be a lot of things, Wyatt Drummond, but *I'm* no coward. Not anymore.

I can't say the same for you, though. You can't even be a man when you murder me. You have to use an impersonal gun. A real man would use his hands."

He moved in front of me, lowering the gun in his right hand as he lifted his left hand to backhand my face. But I was prepared. As he swung, I rushed forward and raised the letter opener, aiming for his chest. Only his arm deflected mine. I missed, instead plunging the letter opener deep into the side of his neck.

I released and gasped, taking a backward step away from him.

He stared down at me in horror. He lifted his right hand as an afterthought, but the movement was slow and labored. I shoved his hand before the gun discharged then fell to the floor.

I ran behind my desk and pressed the button on my phone for security. "I need the police and an ambulance in my office. *Now*."

Wyatt fell backward onto the chair, holding his hand up to his neck. Blood slowly spurted out between the blade and his skin.

I started to cry. Even after all the shit he'd pulled, I'd still hoped there was good in him somewhere.

I considered grabbing something to stop the blood flow, but I didn't know if I'd make it worse, and part of me was terrified he'd still try to hurt me.

I watched him for what seemed like forever, but it was probably less than a minute when I heard fast-running footsteps pounding outside my office.

"Carly?" I heard Marco's panicked shout from down the hall. He must have heard from security. What was he doing here?

"In here!"

He ran through the door seconds later, then horror covered his face when he saw who was in the chair.

"Max," he called out, turning around. "Stay out."

But of course, Max didn't listen, pushing past his best friend and rushing to Wyatt's side.

"I didn't mean to stab him there," I said, starting to cry. "He was going to shoot me and..."

Max dropped to his knees next to his brother. "Wyatt, you stupid shithead," he said through his tears. "*Why?*"

Wyatt didn't respond. He just stared at his brother.

I ran for Marco, letting him engulf me in his arms.

"I swear to God, I'm never letting you out of my sight again," he said, burying his face in my hair.

"I'm good with that," I said as I began to shake. "What are you doing here?"

"I knew you'd forget to leave, so I brought Max to show him the office."

"You were a crap brother," Max choked out, "but you were *my* brother. And I always loved you. I still do."

Wyatt's hand flopped on his leg, and Max took it, squeezing tight. "Fuck you, Wyatt," he said without heat. "How am I supposed to do this without you?"

Security rushed through the door and hurried over to Wyatt and Max, but Max held onto his brother's hand until the blood stopped flowing from Wyatt's neck. Then he got up and walked over to us, looking like a ghost.

"I'm sorry, Max," I said, starting to sob in earnest, "I'm so sorry. I didn't want to do it, but he—"

Then, to my surprise, he wrapped his arms around me. "I'm the one who's sorry. I love you, Carly. I'm sorry he did this to you." Max broke down into tears. "I don't deserve to be part of your lives after what my family's done."

"Max," Marco choked out, placing a hand on his back.

"I'm all alone now," Max sobbed. "They've all left me."

"You're not alone," I said, holding him tight. "You can't get rid of me that easily," I said through my tears. "And Marco, well, you're stuck with him for life."

"You know it," Marco said, hugging us both.

The police showed up soon afterward, and between the security camera in my office and Wyatt's history, they said they had no qualms about labeling his death self-defense.

I struggled to absolve my guilt so easily.

After Wyatt's body was removed and we'd all given statements, Max looked more lost than ever.

"So what now?" he asked.

He'd lost his entire family, not that they'd been much of a family to begin with, but I knew what it felt like to find yourself all alone.

"Now we go home," I said, looking up at Marco and basking in the love on his face. "We all go home and start again."

NOT READY TO SAY GOODBYE?

Read the *Keeper of Secrets* Bonus Chapter! Carly goes back to Henryetta to see Rose, Joe, Jed, Neely Kate and more! This WILL contain spoilers for the Rose Gardner series.

You can access it by using this link or scanning this QR code:

LITTLE GIRL VANISHED

HARPER ADAMS MYSTERY #1

Get ready for a new series featuring several characters from the Rose Gardner world coming in Spring 2023!

Disgraced, ex-detective Harper Adams is perfectly content wallowing at rock bottom. She's lost her job, house, and reputation, and it only gets worse when she's forced to move back home to Jackson Creek, Arkansas, the one place she swore she'd never return.

But everything changes when she gets a chilling phone call. A childhood friend's daughter is missing...and it's eerily reminiscent of Harper's sister's kidnapping and murder two decades ago.

Not her business.

Except with the police are fumbling the case at every move, Harper realizes she might be the only one to save the girl.

And she might need the help of a former crime boss to do it.

Scan the code for more information!

WANT TO CONNECT?

Join my Facebook group! I post teasers, do live videos, and even offer giveaways!

https://www.facebook.com/groups/DGSreaders

Be sure to join my mailing list to keep up with the latest about new releases, sales, and book signings!

ABOUT THE AUTHOR

Denise Grover Swank was born in Kansas City, Missouri and lived in the area until she was nineteen. Then she became a nomad, living in five cities, four states and ten houses over the course of ten years before she moved back to her roots. She speaks English and smattering of Spanish and Chinese which she learned through an intensive Nick Jr. immersion period. Her hobbies include witty Facebook comments (in own her mind) and dancing in her kitchen with her children. (Quite badly if you believe her offspring.) Hidden talents include the gift of justification and the ability to drink massive amounts of caffeine and still fall asleep within two minutes. Her lack of the sense of smell allows her to perform many unspeakable tasks. She has six children and hasn't lost her sanity. Or so she leads you to believe.

For urban fantasy: dgswank.com
For mystery and romance: denisegroverswank.com

Printed in Great Britain
by Amazon

17919884R00174